W9-AAT-521

Constructed Climates

Constructed Climates

A PRIMER ON URBAN ENVIRONMENTS

William G. Wilson

THE UNIVERSITY OF CHICAGO PRESS • CHICAGO AND LONDON

William G. Wilson is associate professor of biology at Duke University. After earning a Ph.D. in physics at the University of Hawai'i at Manoa, postdoctoral studies took him to the University of Calgary and the University of California at Santa Barbara, where he turned toward problems in theoretical evolutionary ecology. His current focus is on environmental issues surrounding human-altered landscapes.

The University of Chicago Press, Chicago 60637
The University of Chicago Press, Ltd., London

Printed in the United States of America

20 19 18 17 16 15 14 13 12 11   1 2 3 4 5

ISBN-13: 978-0-226-90145-9 (cloth)
ISBN-13: 978-0-226-90146-6 (paper)
ISBN-10: 0-226-90145-9 (cloth)
ISBN-10: 0-226-90146-7 (paper)

Library of Congress Cataloging-in-Publication Data

Wilson, William G. 1960–
Constructed climates : a primer on urban environments / William G. Wilson.
p. cm.
Includes bibliographical references and index.
ISBN-13: 978-0-226-90145-9 (cloth : alk. paper)
ISBN-10: 0-226-90145-9 (cloth : alk. paper)
ISBN-13: 978-0-226-90146-6 (pbk. : alk. paper)
ISBN-10: 0-226-90146-7 (pbk. : alk. paper) 1. Urban ecology (Biology) 2. Urban climatology.
3. Environmental health. I. Title.
QH541.5.C6W55 2011
577.5′6—dc22

2010021763

⊗ The paper used in this publication meets the minimum requirements of the American National Standard for Information Sciences — Permanence of Paper for Printed Library Materials, ANSI Z39.48-1992.

To Mom and Dad,
who experienced all the
challenges of a family farm

Contents

Preface

Motivation and Goals

As our world becomes increasingly urbanized, an understanding of the context, mechanisms, and consequences of urban environments becomes more important. This issue involves both the climatic environment in and around cities and the remnants of the natural world in cities, or what I'll generally call "urban open space."

Open space means exactly what it sounds like: forests, pastures, and fields in rural areas, and in urban environments, anything ranging from small gardens to large parks. Open space plays important roles in both vast forested watersheds and congested urban areas where even a single tree can brighten someone's day.

Service on a local citizen group, Durham's City and County Open Space and Trails Commission, was my first introduction to the issues behind urban open space. I spent the first couple of years waiting for the commission, as a whole, to be asked important questions by elected officials, but after a while I concluded that nobody was going to tell me or us our role. One part of the commission, the trails committee, had clear and present directions helping make citizen demands for trails throughout the region a reality, but the open space committee had much less citizen input or appreciation.

Our little open space committee clearly needed to figure out something to work towards. Taking stock of the situation, we could see that despite large-scale development, many parts of Durham County remain quite rural with plenty of privately owned open spaces, and dedicated staff seeking long-term preservation. Many local conservation groups protect natural areas quite forcefully, and there exist a few watershed protections (under continuous threat from development) given the recognized drinking water requirements throughout the region. Instead of focusing on rural areas, we started pushing the idea of Urban Open Space with the goal of protecting and developing nature and vegetation within the city. We boldly put forth a resolution asking for city and county support to direct the Planning Department to start work on an Urban Open Space plan with grand ideas of an even greener city.

After some time, I realized that I, for one, couldn't really provide a good cost-benefit argument for preserving and creating natural spaces in the city. Why should the city and county spend scarce money planting trees when schools need more resources, roads and sewers need maintenance, and people suffer harm at the hands of others? I couldn't really answer those concerns or give insight into those trade-offs beyond providing a vague idealistic image. Seeking an answer to that question turned out to be a wide-ranging study, and addressing it requires a broad synthesis of many environmental topics and human issues. Many of these issues involve the relative values of cleaner water versus smarter children versus calmer citizens versus higher taxes. These values lead to choices and compromises that science alone can't make, but science can and must inform elected officials and citizens making the choices. With this book, I intend to summarize that science and provide the synthesis needed to inform a broad range of interested students and citizens while they carry out these difficult compromises.

Content and Structure

Although many people presently support planting trees in the name of carbon sequestration and sustainability, city natural spaces also involve economics, ecology, social aspects, and air and water quality. This book summarizes my study of urban environments, the role vegetation and trees play, and costs and benefits humans experience from urban open space.

There are a number of goods, but what resonates most with me concerns the issues of emotional health. Near the end of the book I show that low-income areas with rental housing have less vegetation, and it may be that by enhancing low-income citizens' lives through the planting of trees and shrubs — an active demonstration of a long-term investment in people's lives — a city can promote positive social interactions right where they're needed most. Of course, it's hard to believe that a lack of trees drives the conditions faced by low-income residents, but some evidence suggests that urban vegetation alleviates social ills. It makes sense. I truly enjoy gardening and love seeing plants regenerate every spring, a love probably instilled in me from growing up on a farm, watching my parents plant things for their future (and mine). As people of the cities become more accustomed to instant access and communication, and more detached from the delayed gratification needed to spend time planting something now with a payoff months or years off in the future, we face a greater challenge in making sure the cities we live in will have 50-year-old trees 60 years from now.

Another important function involves stormwater runoff, a topic I've only barely touched upon here. In my part of the country, rain falls on impervious surfaces and washes off pollutants. This stormwater runoff brings those substances with it into streams where organisms live and into reservoirs that serve people with drinking water: cleaner runoff, cleaner city water. Urban open space can play a role in purifying that water.

A few quick words about the book's approach and structure: When approaching any research question, one faces balancing depth and breadth. Each brief topic I examine warranted independent theses and books, evidenced by the cited primary literature, and the breadth covered here sacrifices discussing many known details. The plots and results I show often reflect just one small part of a greater whole, leaving out the various nuances about a particular topic that an entire publication or research discipline considers. I bring in just enough material to fill a several-hundred-page book that, I believe, introduces the many topics concerning this synthetic area of urban environments. Nevertheless, I hope my summary accurately reflects the author's relevant conclusions, or at least some small part of their study.

Graphs and plots provide the best summary of scientific results, and I use them liberally. Each of the plots I show comes from one or more peer-reviewed or agency-published publication,* and, on occasion, new and unpublished data. I've also shown a clear preference for the underlying data rather than statistical analyses. Some important studies present the results of formal and complicated statistics without showing a single plot of underlying data. It is certainly correct that scientists must perform statistical analyses, but I find the data even more appealing, with the statistical measures backing up and supplementing the data, not replacing them. Although that's my preference, showing only data is not a rule, and situations exist when data plots become unworkable.

Probably the most apparent structural feature I've introduced and used throughout is the "two-page" format, with plots, tables, and diagrams on left-hand, even-numbered pages, and the relevant explanatory text on the facing right-hand pages. As I stated above, my goal is covering the breadth of urban environment topics in a limited space. This format constrains how deeply I examine each topic, though to relax this constraint I use endnotes extremely liberally for citations,

* In many cases I used a wonderful little open-source JAVA program, *Plot Digitizer*, freely available from *plotdigitizer.sourceforge.net*, to extract data from published graphs and replot them here.

comments, and further details. In some ways the format puts concepts on an equal footing: Some topics have many studies, and others have very few, a variation that reflects differences between ease of study, commercial relevance, environmental importance, or human-health motivations. Those differences might not truly represent relative levels of importance to urban environments, and my hope was to cover more equally what I believe are important aspects of urban environments. Further, I think of the rigidly enforced two-page constraint as demanding a prioritization of information rather than expanding sections here and there as more content becomes available.

The format also reflects my visual nature, or my preference for data plots because the data speak for themselves much of the time. Why use squishy words to talk about a concept when the data show it more clearly? Granted, not everyone has such an inclination. For readers uncomfortable with reading graphs I've added an Appendix covering the basic concepts and ideas contained within.

In addition to plots, I provide various calculations throughout the book, many of them rough ones called order-of-magnitude calculations, designed primarily to see if something makes sense, an approach that checks whether different numbers from different concepts generally agree or disagree. This approach originates with my physics training and serves scientists well under many situations. Furthermore, I have struggled with the very odd combinations of measurement units, for example, the volume of stormwater from inches of precipitation per square meter. I've tried to make the units work with our U.S. system, so no matter what one does, that attempt becomes awkward.

In any event, I'll provide comments, clarifications, downloads, corrections, and updates accessible from my Web page, biology.duke.edu/wilson.

Acknowledgments

There are many people to thank for their help in many different ways: all my fellow Durham Open Space and Trails (DOST) commission members, in particular the open space committee members, and our hard-working staff, especially Jane Korest and Helen Youngblood. It was this connection that ultimately motivated my examination of urban environments.

I greatly appreciate Joe Sexton's efforts, who provided critically important and highly motivating data on the temperature and canopy profiles across Durham County. I'm also grateful to Rob Schick, who coupled canopy information and Census data, which prompted an understanding of environmental inequities.

Many thanks also to the many people I've corresponded with over these several years, including Bob Bornstein, Russ Bowen, John Cox, Grady Dixon, Chris Ellis, Theresa Fisher of Vaisala, Inc., Guido Franco, Chris Geron, Noor Gillani, Peter Groffman, Morgan Grove, Alex Guenther, Pat Halpin, Andy Hansen, Sharon Harlan, Nik Heynen, Alex Johnson, Sujay Kaushal, Rachel Kaplan, Chip Knappenberger, Frances Kuo, Gil Liu, Jim MacDonald, Dan McShea, Bill Morris, Jamie Pearce, Jennifer Peel, Tom Peterson, Chantal Reid, Carl Salk, Sandy Sillman, Jonathan Silverman, Tony Stallins, Will Stefanov, Duane Therriault of Durham GIS, Todd Twigg, Dan Walters, Jeff Wilson, Jennifer Wolch, and Jun Ying. Beyond a doubt I have forgotten important names in this list: my apologies. Numerous anonymous reviewers also put in greatly appreciated efforts that made the book better. I also benefited from the fortitude of my Spring 2007 Ecosystem Services seminar attendees, as well as students in my Cities and Trees classes. Finally, a book like this one could never have been written without the principles of academic freedom that the tenure system provides.

William G. Wilson
Durham, NC
2/4/2010

Constructed Climates

Chapter 1

Cities and Nature

Humans altered the biosphere, greatly expanded their population, and now feel the effects of those alterations. A hunter–gatherer lifestyle gave way to an agrarian one, and an agrarian lifestyle gave way to an urban lifestyle. Urban dwellers exist more detached from the biosphere and from the environment that shapes and interacts with it. In this book I review the ecological, environmental, and sociological features of urban life and a world increasingly changed by its human occupants.

The human population exploded in recent centuries, reaching numbers 60 times greater than the population 2,000 years ago and sixfold more than just 150 years ago, bringing with it serious environmental challenges. In this chapter I put human population densities into the context of other creatures and explain several important ecological concepts that show the stark immensity of our present population. I look at people's use of land and water, and how human densities and resources connect to important general ecological principles. Reasonable estimates show that our global population exceeds "natural" levels by up to 400-fold. If we were just another species, then, given our body size there ought to be about one person per square kilometer. However, suburban and urban areas of the United States have densities of 1,000 to 10,000 people per square kilometer.

All these people, mostly concentrated into cities, put tremendous stresses on our natural resources. Water and fertilizers represent two fundamentally important aspects of sustaining a large human population because both factors relate to important ecological features, including evapotranspiration and net primary productivity. These two features drive many aspects of nature, an important one being biodiversity, meaning the number or richness of species within broad groupings of organisms. For humans, evapotranspiration is important to demands of growing food, but I also demonstrate our immense dependence on producing nitrogen fertilizers using fossil fuel energy. While fertilizer greatly amplified food production and supported more people, along with increasing technologies, it meant an increased agricultural efficiency. That change in efficiency affected the economic viability of small family farms, and as these businesses collapsed, it pushed a tremendous land-use change toward urbanization.

An important feature of increased urbanization is that when precipitation falls on impervious surfaces, the water makes its way through constructed stormwater systems to urban streams. As a result, urban streamwater quality suffers, killing the sensitive organisms living in urban streams. These consequences correlate directly to the amount of impervious surface contained in a watershed. I refer to many examples from my city, Durham, North Carolina, and like most cities, Durham's stormwater system uses urban streams as above-ground stormwater pipes, and the stormwater flows into drinking water reservoirs. As a result, decreased water quality in urban streams correlates with various measures of increasing urbanization.

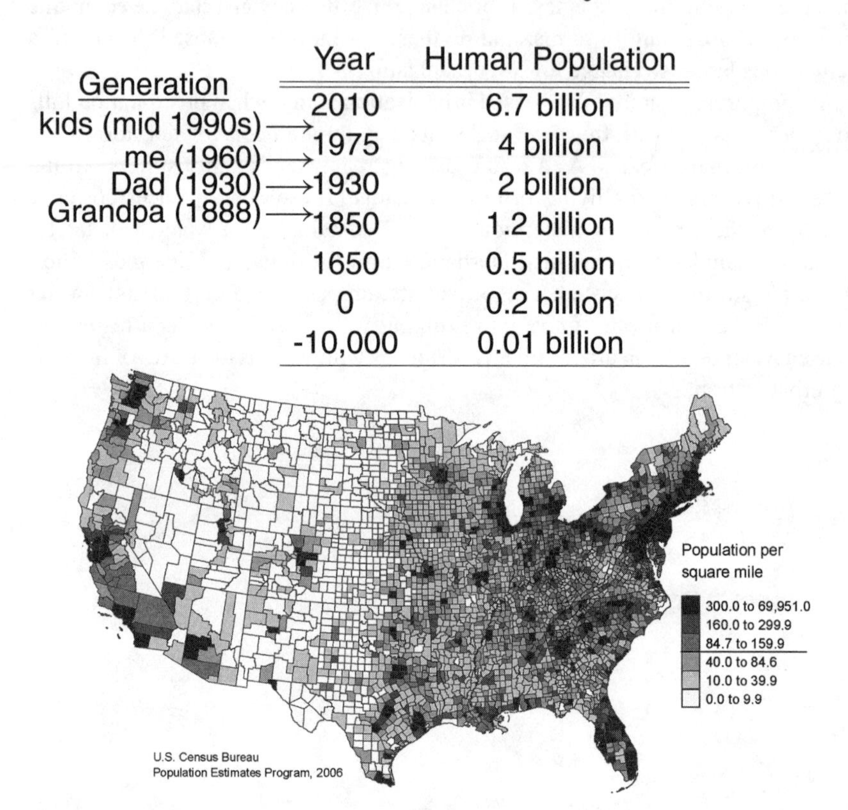

Human population increased sixfold over the last century.

Generation	Year	Human Population
kids (mid 1990s) →	2010	6.7 billion
me (1960) →	1975	4 billion
Dad (1930) →	1930	2 billion
Grandpa (1888) →	1850	1.2 billion
	1650	0.5 billion
	0	0.2 billion
	-10,000	0.01 billion

Figure 1.1: The table shows global population numbers exceeding estimated "natural" levels by as much as 200-fold, increasing more than *sixfold* since just the mid-1800s. The year -10,000 simply means pre-agrarian. Across the United States, the U.S. Census map shows county population densities varying from 0.1 people per square mile (Loving County, Texas) to more than 70,000 people/mi^2 (New York County, New York).

A book on cities and nature needs the context of just how many people exist in reference to the expectation of "natural" levels because the recent human population increase underlies many environmental features and concerns we experience today. More people means more cities, less "nature," and a greater role for urban environmental conditions that affect Earth in myriad ways, local and global, with consequences lasting well into the future.

I find current population numbers even more sobering when compared to how many people existed long ago. In the table at the top, I put our population explosion into a personal context: My oldest grandparent was born in 1888 when there were roughly 1 to 2 billion people in the world. My dad was born in 1930 when there were about 2 billion people. I was born in 1960 when there were about 3 billion people. My oldest child was born in the mid-1990s when there were about 5 to 6 billion people. Now there are about 6.7 billion people.[1] Human population increased sixfold in 120 years, just four generations! (See Figure 1.1.)

Going back further in time, we find that human population increased by about a factor of 60 over the last 2,000 years. If you accept the idea that humans were once just another species, comparisons with other organisms yield around 15 to 150 million people for our "natural" population (see Figure 1.4), which we now exceed by as much as 200-fold.

Let me put population growth another way: Up until two millennia ago, during nearly all of humanity's existence, our population tripled. During the last 200 years, our population tripled and then tripled again. That's a lot of people added to Earth in just the last 100 years or so.

How densely are people packed? Earth's land surface measures 150 million square kilometers, and we have 6.7 billion people: simple division gives 40 people per square kilometer (km), or 104 people per square mile.[2] Across the United States and as seen in Figure 1.1, populations range from very low densities, like the 0.1 people/mile^2 (0.039 people/km^2) in Loving County, Texas, and several counties in Alaska, to a very high density in New York County, New York, with 70,700 people/mile^2 (27,300 people/km^2) over its 22.8 square miles.

All these people need resources, and so we'll next examine food production and land-use change.

Water, warmth, and light make plants grow.

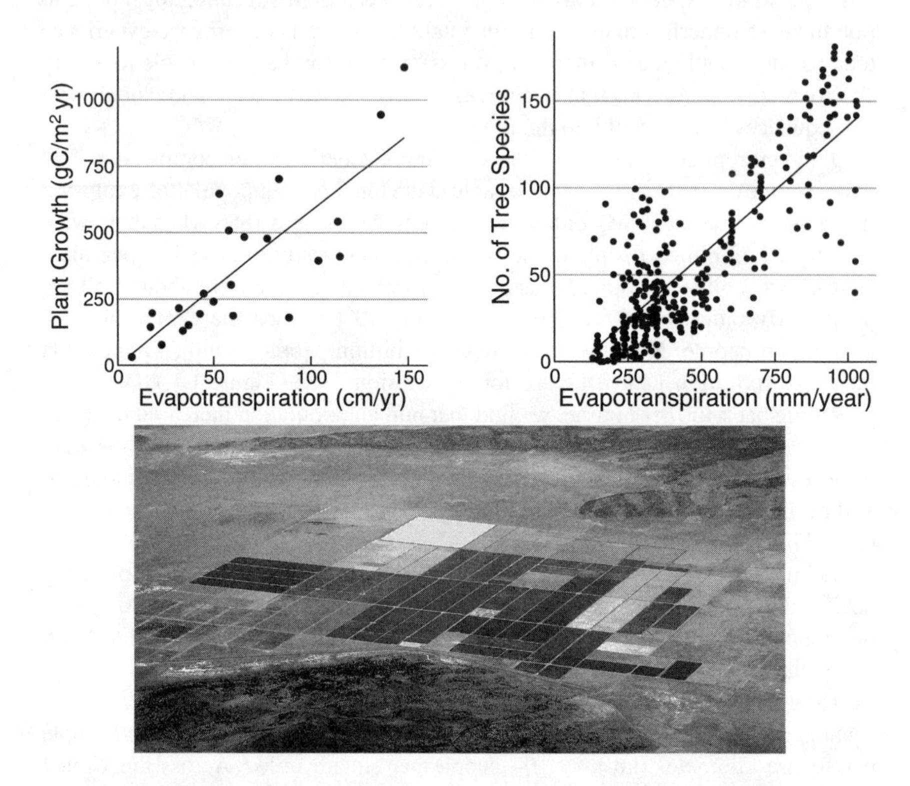

Figure 1.2: The top left graph shows plant growth (called net primary productivity) across 23 ecological areas with different levels of evapotranspiration (data from Cleveland et al. 1999). My desert photograph shows green areas living up to their high *potential* evapotranspiration through irrigation, whereas the surrounding brown area has a much lower *actual* evapotranspiration strongly limited by low precipitation. Interestingly, the top right plot shows that the number of tree species also increases with an area's evapotranspiration (after Currie 1991).

Humans need to eat food, food ultimately requires plant growth, and plant growth requires water, warmth, light, and nutrients. It really comes down to evaporating water. Ecologists sweep all the biologically relevant ways of turning liquid water into vapor under the term *evapotranspiration*, but only two really important ways catch my interest for the questions at hand: simple evaporation of liquid water, which accelerates when water is heated, and transpiration, which takes place when plants use and lose water while growing.

Two main ingredients limit evapotranspiration: heat and water. I took the desert photograph shown in Figure 1.2 out of a plane flying over the western United States, and, in a wonderful way, it shows the difference between potential evapotranspiration and actual evapotranspiration. A location's potential evapotranspiration measures how much liquid *could* change into vapor if only an unlimited water supply existed. Assuming infinite water means that heat ultimately limits potential evapotranspiration: How much water can a location's heat evaporate and transpire? In the Arctic, for example, there's plenty of (frozen) water, and if only there were more heat there'd be more evapotranspiration. Actual evapotranspiration, in contrast, tells how much evapotranspiration *actually* takes place in a spot given both its heat and water. A large difference between potential evapotranspiration and actual evapotranspiration of the desert motivates the farmer irrigating the fields.[3]

The upper left plot turns my photograph into numbers,[4] showing how the production of plant material increases as evapotranspiration increases[5] across 23 vastly different ecological areas ranging from deserts to rainforests.[6] My photo shows this graph's most extreme points in just one spot: the dry extreme by nature, another extreme made wet by irrigation.

For ecologists, plant growth means the uptake of carbon from the atmosphere, while appreciating that plants both take up carbon through photosynthesis and release carbon through respiration. We call the balance — uptake minus release — *net primary productivity*, or NPP for short. This plot shows that the net amount of carbon absorbed by plants depends on how much warm water they have available. Annual NPP totaled across the continental United States amounts to 3.4×10^{12} grams of carbon distributed over the United States' 7.9 million km^2, giving an average plant growth of 430 gC/m^2/year.[7]

Biomass production excites people, historically for food and more recently for energy. However, packaging biomass into species[8] excites botanists and ecologists even more. Along with increased plant biomass production, the upper right plot of Figure 1.2 shows that increasing evapotranspiration also correlates with an increasing number of tree species.[9] Through some unclear mechanism, the process of natural selection yielded more species where greater plant growth occurs.

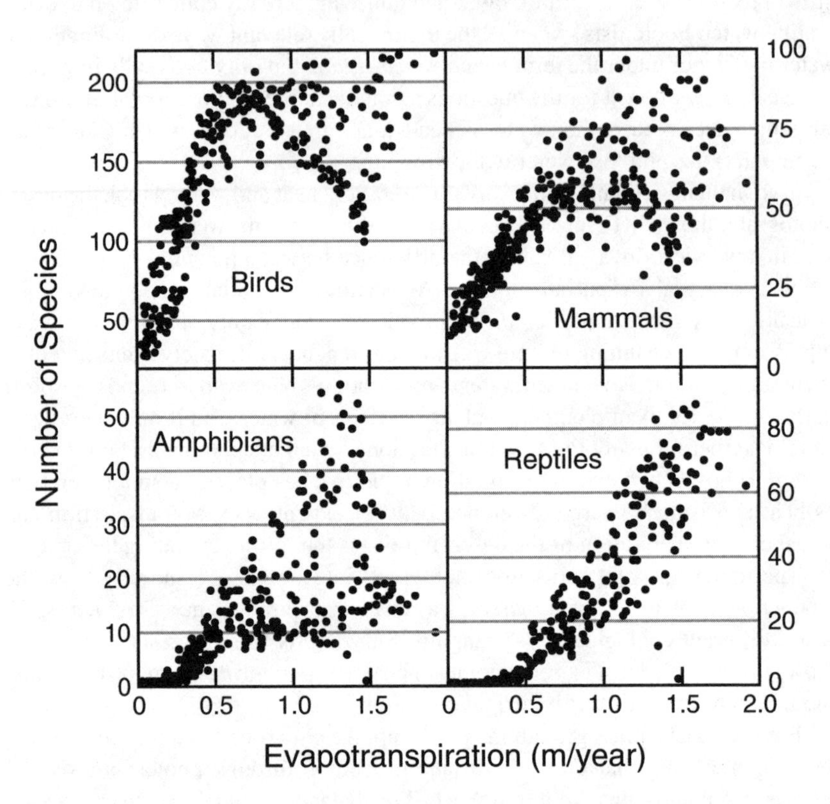

Figure 1.3: Along with the number of tree species in Figure 1.2, the number of animal species also increases with evapotranspiration. The overall pattern repeats for diverse groupings such as birds, mammals, amphibians, and reptiles across the globe (after Currie 1991). Each point represents a different location, and species counts come from distribution maps found in field guides used by naturalists.

The number of species in a given ecological or evolutionary grouping is called its *species richness*, and we just saw that tree species richness increases with evapotranspiration. So too does the species richness of birds, mammals, amphibians, and reptiles (Figure 1.3). In other words, add heat, water, and sunlight, and out comes lots of plant growth with lots of species in each general grouping of organisms. As with body size scaling (Figure 1.4), the precise reasons behind these observations aren't well understood — yet there's the empirical observation. The data come from compendia of species range maps, perhaps one of the field guides that might be sitting on your bookshelf. An ecologist digitized and overlaid each map, creating a spatial map of species richness for each taxon of organism. Combining that information with independent evapotranspiration data generated these plots.[10] Details differ between the taxa, such as the amount of scatter and saturation or downward trends at very high evaporation, but those details seem unimportant overall.

Evapotranspiration clearly drives not only how much plants grow, but also the conditions it promotes, like abundant plant growth, which influences plant and animal biodiversity. According to the study behind these figures, evapotranspiration accounts for more than 90% of the variation in species richness among these organisms! This high value means that in a natural area with minimal human disturbance, all other ecological factors besides evapotranspiration — interesting ecological interactions such as competition, predation, and disease — independently account for less than 10% of the variation behind how many species live there.

Species need evapotranspiration, and evapotranspiration needs water. Water comes from rain, and rain comes from evapotranspiration. Rain enters the ecological loop by soaking into the soil, then being pulled back out by growing plants or evaporation. This key feature has implications for cities. In various places throughout this book, I discuss impervious surfaces such as roofs, roads, buildings, and parking lots. Rainwater can't percolate into impervious surfaces, interrupting this crucial dependence of plant growth and species richness on evapotranspiration. A hypothetical city, half covered by impervious surfaces, might experience half the evapotranspiration it should because stormwater systems divert rainfall directly to streams, ponds, rivers, lakes, and ultimately oceans (see Figure 1.10). This reduction reduces plant growth, sustains fewer species, and results in fewer benefits from vegetation.

Imagining that humans were just another species, we'll now compare our population numbers with those of other organisms.

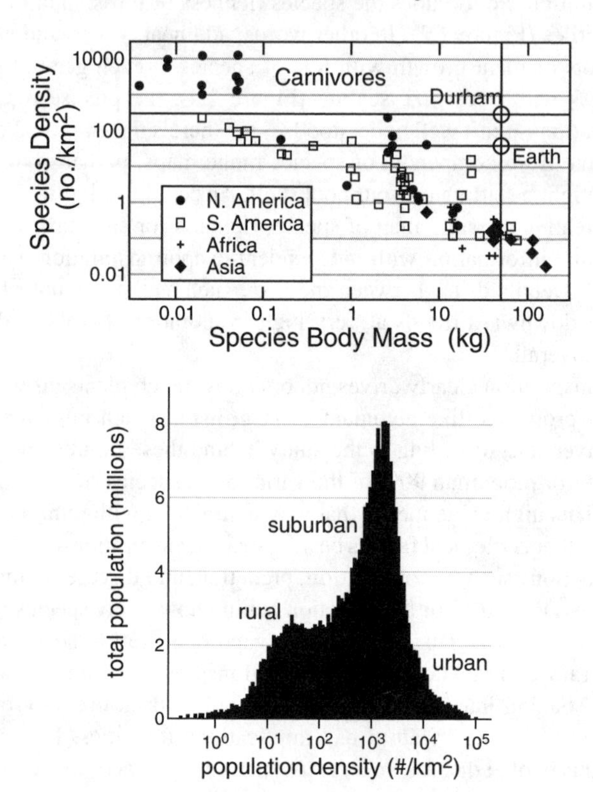

Figure 1.4: Population densities of mammalian carnivores correlate with each species' body mass (after Peters 1983; herbivore plots are similar), meaning the larger the body, the lower the population density. In this plot, each data point marks a different species. Assuming a human body mass of 50 kilograms (110 pounds), prehistoric populations span 0.1 to 1.0 people/km^2. A distribution of 1990 U.S. populations shows that this range matches only the low end of rural densities, and suburban and urban populations exceed it by several orders of magnitude (after Pozzi and Small 2001).

Let's put today's human population into perspective. Long before humans became really clever, our population fell in line with the populations of other organisms. That assumption seems reasonable to me, and it provides an estimate for prehistoric population sizes. The plots shown in Figure 1.4 demonstrate the connection between a carnivorous (meat-eating) species' population density and individual body mass. Mammalian species from different continents fall onto a general trend line, showing that the relation holds across different ecological communities. Species with small individuals have lots of them: many, many mice, but very few elephants. Quite some time ago biologists saw this clear and strong connection between population density and body size, not to mention many other quantities, but still there's no consensus as to why it should be so. Ecologists call this fascinating research area *body size scaling*, and many scaling relations exist for other ecological and physiological quantities.[11] Humans, being mammals, were subjected to whatever biological forces impose this body size scaling, and these plots provide rough limits on a prehistoric population size estimate, as well as a stark indication of how far off the line our tremendous recent population increase has pushed us.

To find our place in the top plot, let's assume a human body mass of a nice, svelte 50 kilograms (kg), which corresponds to a weight of 110 pounds (lb), falls within empirical reason for prehistoric people.[12] Putting this number on these plots provides a rough expectation for our "natural" population density: *One person per 10 square kilometers* if we were strict carnivores, and *one person per square kilometer* if we were strict herbivores (from results of a similar curve). We expect, then, that natural human densities ranged between 0.1 to 1.0 people/km^2 (0.3–3 people/mi^2), wherever conditions were suitable for primitive human habitation. Compare these densities to the modern ones across the United States (Figure 1.1), ranging from 0.04 to 27,000 people/km^2 and shown in the bottom plot of Figure 1.4 for 1990 U.S. populations.[13] Nowadays, rural might be considered anything less than 100 people/km^2, urban more than 10,000 people/km^2, with suburban in between. These numbers exceed humanity's present world-averaged density of 40 people/km^2, which assumes we spread people out over every patch of Earth.

Multiplying the above natural density estimates by Earth's land surface, 150 million km^2, yields the globe's prehistoric human population, somewhere between 15 and 150 million people. If humans were strict carnivores, the global human population would be about 15 million; if strict herbivores, there would be about 150 million humans.[14] Regardless of diet, these natural estimates fall much lower than Earth's 6.7 billion people, a number roughly 40 times greater than for herbivores and 400 times greater than for carnivores.[15] In sum, the human population greatly exceeds any sense of a natural carrying capacity.

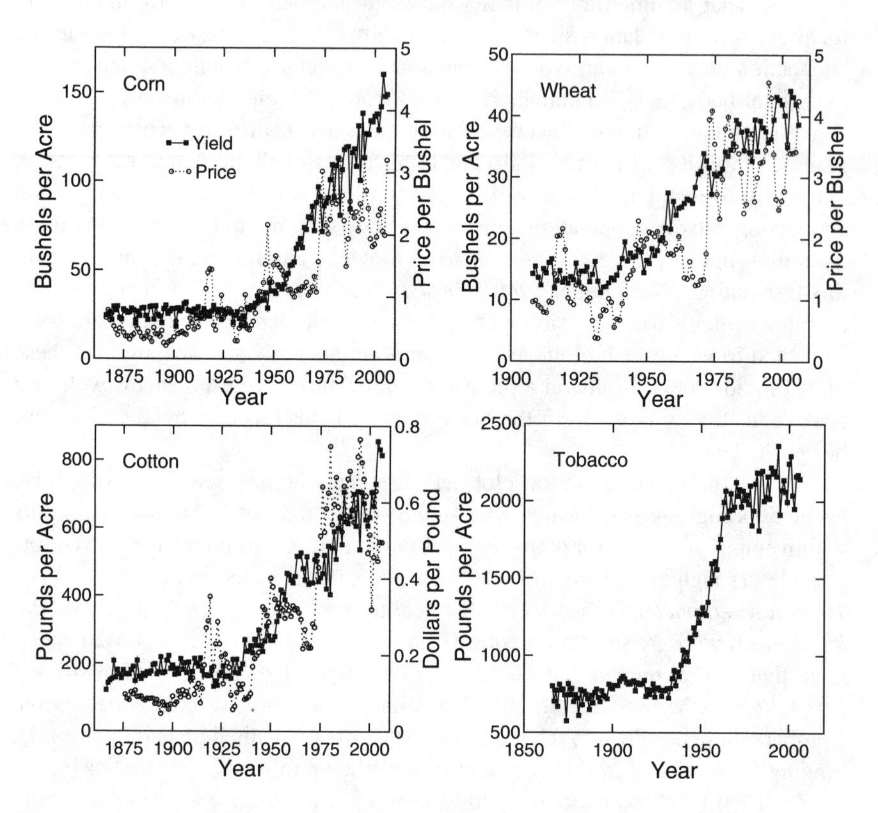

Figure 1.5: These four plots show the yields and prices in the United States for corn, wheat, cotton, and tobacco (without price). Yields per acre (filled squares) increased dramatically, whereas the price per bushel (open circles) exhibits somewhat smaller increases with greater variability (data from www.nass.usda.gov/QuickStats).

More people should mean more food eaten, meaning more farmland needed. Instead, farmers grow crops much more efficiently today than just a few decades ago,[16] with six times higher crop yields. Despite an increased population demanding more food, Figure 1.7 shows that U.S. farmland has held mostly steady over the last 60 years. We can feed all these people precisely because we've modified and engineered a few other species' productivities. Mechanical, environmental, and biological engineering[17] supports our huge population.

I grew up on a farm and can put Figure 1.5 into a personal context. One of my uncles recalled how his Minnesota family traded in their team of workhorses for their first tractor in 1941, right when crop yields take off. In fact, my brothers and I never had a horse while growing up because my late father — a fantastic mechanic — disliked horse chores as a child. The rejection of horse power in favor of fossil fuel power keeps people fed.

The U.S. Department of Agriculture (USDA) estimated that a farm needs $250,000 in annual crop or livestock production to break even (see Figure 1.7), that is, just to have income meet expenses.[18] What does that number mean to city people? Let's assume corn was priced at a historically high $3 a bushel,[19] ignoring very recent, and very transient, off-the-axis prices from the biofuels push, and a great yield of 150 bushels an acre.[20] What sized farm earns a farmer a quarter-million dollar gross income? The answer is more than 550 acres of land producing corn, nearly 2,000 acres producing wheat, or about 400 acres producing cotton. In contrast, Durham County farms average just 100 acres.

Are farmers economically sensible? Consider just the land assets involved in a corn-producing farm. If the farmland is worth $5,000 an acre, then its total value is $2.75 million, and that huge asset, not counting equipment, fuel, and labor, provides only a break-even return. On paper, the United States has millionaire farmers who can't make a living. A farmer could (an investment adviser might say should) sell all his or her land, invest it in mutual funds with a 10% annual return,[21] and make $275,000 per year without backbreaking 16-hour workdays. The advice doesn't strongly depend on these figures: Divide the land price by two and the numbers remain astonishing. I'm surprised we have any farms at all in the United States. Essentially, farmers provide volunteer labor, not to mention a public service. Local, organic, and small-scale farming methods gained popularity recently, but it seems questionable that we can afford to scale back the increased food production efficiency from the last century. Certainly these interests can reenergize small farms and, perhaps, even prompt gardening by city dwellers.

Let's now examine the origins and consequences of these yield gains.

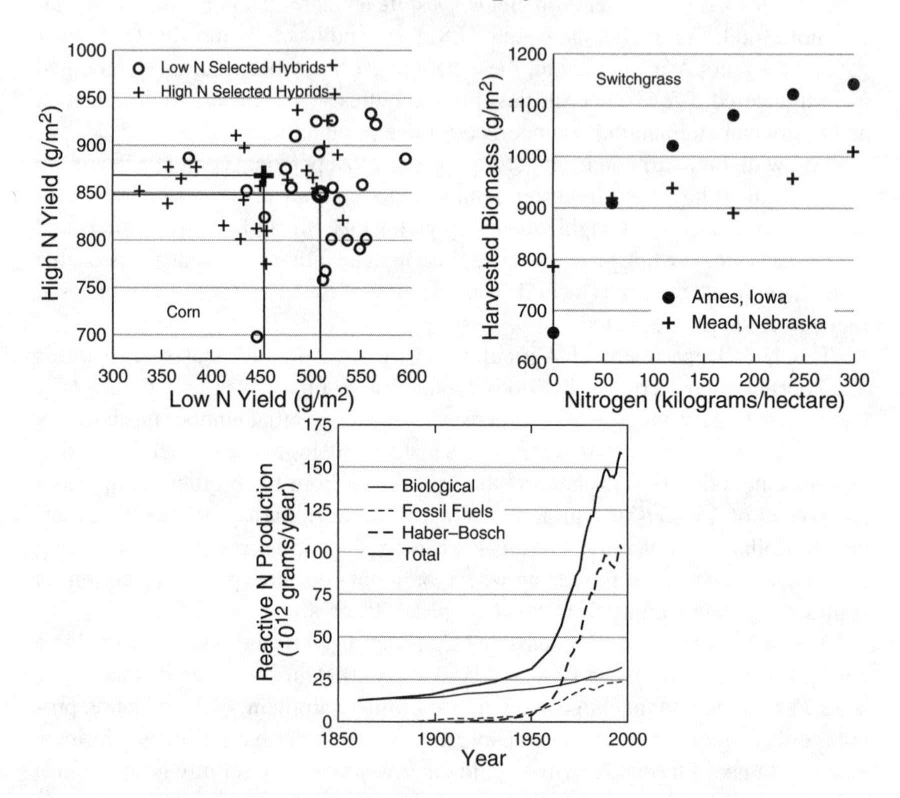

Figure 1.6: Top left plot shows the yields of many corn hybrids, bred for high performance at either low or high nitrogen levels, grown in low and high nitrogen conditions (after Presterl et al. 2002). Selective breeding certainly helps increase yields, but adding nitrogen increased yields from about 500 g/m^2 to about 850 g/m^2. Switchgrass biomass, a much touted biofuel, also increases with applied nitrogen (after Vogel et al. 2002). Artificial production of reactive nitrogen increased greatly over the last century (after Galloway et al. 2003), enabling the large crop yields shown in Figure 1.5.

We owe increased yields, in large part, to nitrogen, and we owe nitrogen to fossil fuels. Plant growth increases with available warm water, but many things can limit plant growth, nitrogen included. Amino acids, the building block of all proteins, have an NH_2 group attached to a carbon atom attached to everything else, so without nitrogen, growing plants have a problem.

Nitrogen limitation happens, in part, by harvesting crops and shipping them to city dwellers, taking available nitrogen away from fields, ultimately getting flushed down toilets rather than entering the local nitrogen cycle.[22] Nitrogen enters the biosphere naturally when fixation takes place in some free-living bacteria, as well as the mutualistic root nodules of some plants, importantly the legumes.

Nitrogen limits biomass production; 2,000 years ago farmers knew that planting legumes reconditions soils.[23] In Figure 1.6 I show two far more recent studies, one involving corn and the other switchgrass, a forage crop recently touted as a biofuel source.[24] Using sets of northern European corn hybrids selected for high yields at low and high nitrogen levels, one plot compares yields at low and high nitrogen levels.[25] Plant breeding helps: Hybrids bred for performance at low nitrogen had 10% greater yields at low nitrogen. Yet, independent of breeding, nitrogen limitation greatly reduces yields. Between these low and high nitrogen fertilizer treatments, available nitrogen slightly more than doubled, and yields increased by some 80%. Likewise, switchgrass crops grown in Ames, Iowa, and Mead, Nebraska, show clear increases in biomass[26] with more applied nitrogen, somewhere around a 50% increase.[27]

Lots of people means growing lots of food, and we just can't get around that fact. Does adding nitrogen make economic sense to farmers? Fertilizer prices for agriculture over the last few years were around $400 per ton (or 900 kg). In the corn experiments the high N treatment hit 200 kg/hectare (179 lb/acre) meaning a ton covers 4.5 hectares (about 11 acres) at a cost of $36 per acre (ignoring application costs).[28] This level provided about 400 g/m^2 yield gain, or 63 bushels/acre. At $3/bushel for corn, the economic benefit reaches $190/acre, a return of about $5 for every $1 spent. A farmer planting 1,000 acres of corn needs that $200,000.

The bottom plot of Figure 1.6 shows total nitrogen use since 1850: Just since 1950 it increased sixfold, as did the crop yields in Figure 1.5. Some nitrogen comes from fossil fuel use — think emissions — and some from planting legumes,[29] but most comes from commercial fertilizers produced via the energy-intensive Haber–Bosch process, invented about 100 years ago.[30] The resulting high-efficiency farming methods prompted the land-use changes shown next.

Urban land use grew as small farms disappeared.

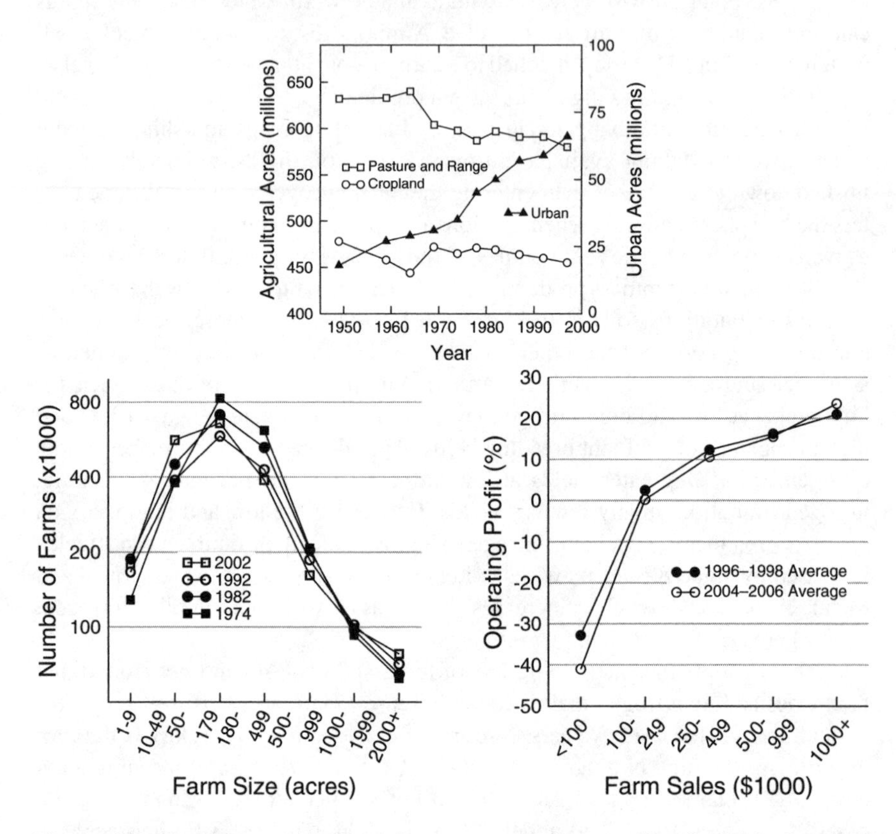

Figure 1.7: Despite a much larger human population needing food, the last half of the twentieth century in the United States saw agricultural land use decreasing by about 75 million acres while urban land use tripled, increasing by about 50 million acres (data from Lubowski et al. 2006). These land-use classifications represent about half of the United States' 2 billion acres (excluding Alaska and Hawaii). The changing farm-size distribution shows that intermediate-sized farms either break apart or merge into larger ones as small farms lose economic viability (data from MacDonald et al. 2006 and Hoppe et al. 2007).

Are cities stealing farmland? The many classifications and changing definitions of land use complicate understanding land-use change.[31] Although the amounts of lost agricultural land roughly equal the gain in urban land, agricultural land is not being converted wholesale into suburbia because forestry land also plays an important part.[32] In any event, only a small part (3%) of the nearly 2 billion acres in the 48 states falls under the urban land-use classification, despite the impression from Figure 1.1.

The curves in Figure 1.7, however, make it clear that the United States gained people and converted land to urban use. Our urban sprawl problems likely originate from the situation underpinning the quarter-million dollar break-even value shown at the bottom right of the figure. The USDA 2002 Census compares farms between 1974 and 2002.[33] During this period the number of farms declined by 10%, from 2.3 million to 2.1 million. The average farm size remained constant at 440 acres, but that constant hides change. Intermediate-sized farms between 180 and 500 acres decreased from 616,000 to 389,000, but small farms less than 50 acres increased from 507,000 to 743,000, and 1000+ acre farms increased from 155,000 to 177,000. Presumably, medium-sized farms failed because of the break-even problem and faced either a merger into larger, economically viable farms or subdivision into hobby farms and urban development.

In the United States, our urban land tripled, agricultural land decreased slightly, and population doubled.[34] All these counterintuitive land-use changes took place because of the agricultural land's increased efficiency through higher yields. Consider, for example, if there were 10 factories creating similar widgets and, suddenly, the productivity of each factory increased tenfold. Widgets would overrun us. Even if the demand for widgets increased, say threefold, the increased supply would make the price of widgets fall. Small factories would lose money and stop production. Similarly, as agricultural yields increased over the last 60 years, small farms became unsustainable. Unlike a factory, a farm can't produce much more than crops and still be called a farm, so farm kids take up new occupations, and the assets — land — get retooled to become subdivisions or forests.

Sadly, the farmland's value represents the farmer's 401(k) or 403(b) retirement plan, a source of health care money, and the only way to pay for potential nursing home care. What choice do retiring farmers have? They can't afford to give the land to their children, and their children can't afford to buy it from them as farmland. As farms transform into suburbs, children grow up without playing in the dirt, climbing trees, watching ants and bugs and worms, learning mechanical reasoning, and learning the delayed gratification resulting from planting a seed in the spring and eating an ear of corn in the fall (see Figure 5.10).

Cities change ecological communities.

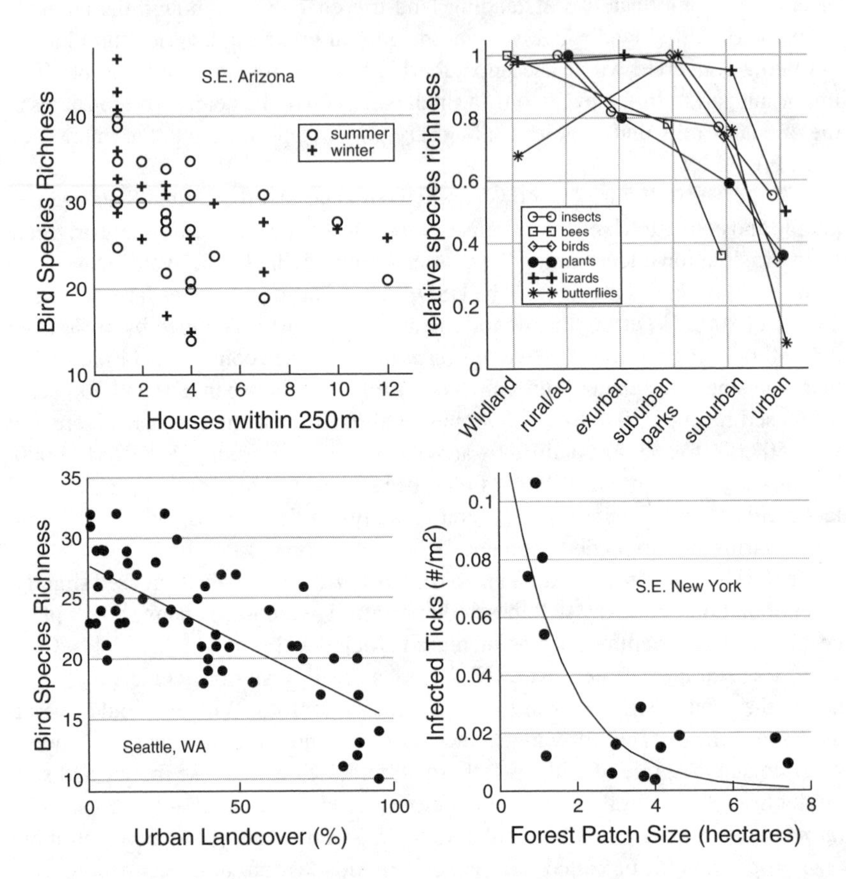

Figure 1.8: Bird species richness decreases with the number of houses in southeastern Arizona (after Bock et al. 2008). Relative species richness of many organisms, sampled over several regions of the United States, decrease with increasing urban land use (after Hansen et al. 2005). In Seattle, Washington, bird species richness decreases drastically with increasing urban landcover over 1 km^2 areas (after Donnelly and Marzluff 2006). Habitat fragmentation associated with urbanization creates small forest patches, which hold a greater abundance of ticks infected with the bacterium causing Lyme disease (after Allan et al. 2003).

Land-use change affects animals and plants, too, and, not surprisingly, denser cities harm ecological communities more (Figure 1.8). Exurban areas in Sonoita Valley in southeastern Arizona are sparsely populated "suburbs" with house densities averaging one house per 5.2 hectares (about 13 acres) with small commercial centers.[35] In these areas, researchers counted birds on 48 transects covering 480 square kilometers, with each transect sampled 29 times, roughly balanced between winter and summer. The upper left graph of the figure plots bird species counted against the number of homes within 250 meters of the transect's midpoint. Interestingly, more bird species were found in exurban housing developments than on undeveloped ranch land, and, surprisingly, more birds were found on grazed than ungrazed lands. Water availability likely explained both of the patterns: It takes just a few ranch houses to provide water to birds, and beyond that, birds dislike more houses.

At top right, a compilation of studies across the United States indicate similar patterns across organisms ranging from insects to plants to lizards. Generally, "wild" places and rural areas show the greatest species diversity, while urban areas are relatively depauperate.[36] Overall, highly urban areas possess just 10 to 50% of the species found in the most species-rich areas.

In Seattle, Washington, a variety of sites were chosen along a gradient of urban landcover, averaged over more than 50 distinct square kilometer quadrats.[37] Birders performed bird species counts at these sites and found a remarkable decrease in species richness, shown at bottom left, from about 25–30 in forested sites, down to 15–20 species in mostly urban areas.

An even more complicated scenario exists with ticks and Lyme disease, shown at bottom right. The conversion of land into housing developments leads to forest fragmentation, and fragmentation causes all sorts of upset ecological interactions. Lyme disease in people results from infection with *Borrelia burgdorferi*, a bacterium transmitted by the tick, *Ixodes scapularis*. This tick loves mammals, birds, and reptiles, with its juvenile stages having been reported on more than 100 different host species.[38] Juvenile (nymphal) ticks get infected by the bacterium when they take a blood meal from an infected host individual, but host species differ in how likely the bacterium gets transmitted to the baby tick.[39] White-footed mice make great hosts from the bacterium's perspective, transmitting the bacterium to the tick around 6 times out of 10 blood meals. Habitat fragmentation comes in because the white-footed mouse wins the ecological competition against other species in small forest patches. The mouse wins, the tick probably doesn't mind, the bacterium certainly wins, and urbanites get Lyme disease.

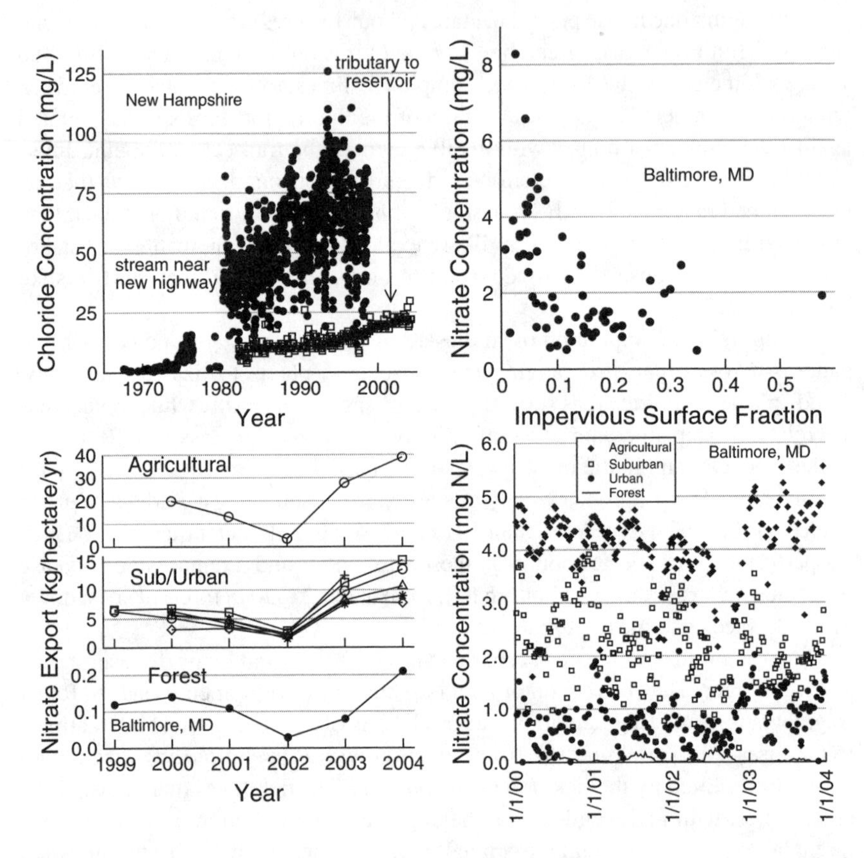

Figure 1.9: At top left, rural stream salinity (chloride concentration) increased after the 1970 interstate highway construction in the Hubbard Brook Valley of New Hampshire's White Mountains (filled circles), and long-term salinization of a tributary (open squares) to one of Baltimore's drinking water sources (after Kaushal et al. 2005). At top right, streams near agricultural areas (low impervious surface fractions) around Baltimore, Maryland, have high nitrate concentrations, leveling off in urban areas. These levels reflect nitrate export levels in agricultural and urban areas, shown at bottom left, which far exceed forest levels (after Kaushal et al. 2008). Weekly sampling shows great variability, though clearly indicating levels much higher than forests (after Groffman et al. 2004).

Land-use change also affects streams, rivers, and lakes. When thunderstorms and lightning strike (see Figures 2.4 and 2.6), impervious surfaces bring up the issue of where rain goes when it falls. Results from the northeastern United States demonstrate a couple of problems with stormwater runoff. The top left plot of Figure 1.9 shows chloride (salt) concentrations in two rural streams alongside roads. Open squares depict salt concentrations in a New Hampshire stream flowing into one of Baltimore, Maryland's drinking water reservoirs.[40] Over a period of 20-some years the concentration nearly doubled. Filled circles (with an inconvenient data gap) show the change in salt concentration of a stream near a highway constructed in the 1970s. Chloride comes primarily from the salt used to de-ice roads and parking lots. Along with the stormwater runoff, salt ends up in streams that empty into reservoirs, lakes, and aquifers. I think of stormwater runoff as parking lot wash-and-rinse water, and in the end, if streams are polluted, downstream reservoirs are also polluted. We drink what we spray on our roads.[41]

The top right and bottom graphs show the connection between nitrogen and land use in the Baltimore area of Maryland.[42] True, agriculture adds nitrogen to streams: Farmers apply roughly 60 kg of nitrogen per hectare per year (recall the importance of nitrogen fertilizers to crop yields in Figure 1.6), but since I depend on food produced in rural areas, and want farmers to survive economically, I'd be rather hypocritical being too upset by their use of nitrogen.[43] On the other hand, suburbanites in these watersheds apply an average of 14.4 kg/hectare/yr to their lawns, which, some might argue, serves no useful purpose. Unfortunately for all of us, not all of this applied fertilizer stays where it should to increase yields and beautify lawns. Some of it runs down ditches and into streams. In the top right plot, high nitrate concentrations in streams occur in areas with lower fractions of impervious surfaces, reflecting runoff from agricultural land. However, the levels remain high for higher impervious surface fractions due to lawn applications.

Nitrogen export differs between land-use types. At bottom left we see that agricultural areas export around 20 kgN/hectare/yr, urban and suburban areas export 5–10 kgN/hectare/yr, while forests export very little. Note the marked export decrease during the 2002 drought: no rain, no runoff. Weekly sampling results, displayed at bottom right, shows the inherent variability in these pollutant levels present in agricultural, suburban, and urban streams.[44] Only forested areas show little variation around essentially no concentration!

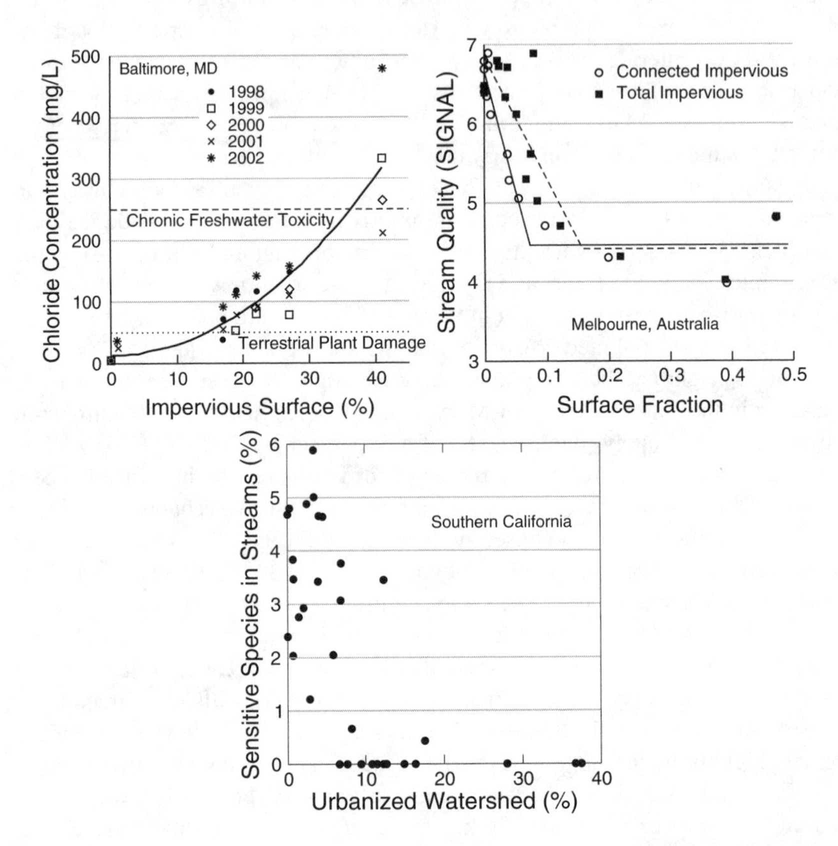

Figure 1.10: The top left plot shows stream toxicity increasing with impervious surface fraction. In comparison to rural streams, suburban and urban streams have chloride concentrations ranging from 200 to 500 mg/L (after Kaushal et al. 2005). At top right, surveys of aquatic insects measure the ecological health of streams in Melbourne, Australia, with the result that stream health seriously degrades with just 10–20% impervious surface fractions (after Walsh et al. 2005b). Impervious surfaces with drainage directly connected to streams, having the least filtration opportunities, harms aquatic life the most. Streams in Southern California show that as little as 8% watershed urbanization causes the loss of sensitive aquatic insects (after Riley et al. 2005).

The examples presented in Figure 1.10 put streamwater quality into a broader context of ecological health. First, at top left, salt concentrations in Baltimore area streams increase with a watershed's impermeable surface fraction. Essentially, more parking lots means worse water. These concentrations exceed the toxicity levels defined by Environment Canada and the U.S. Environmental Protection Agency (EPA) for healthy streams, and far exceed levels tolerated by terrestrial plants, implying bad things for streamside vegetation. Other results I haven't shown for these streams indicate salt concentrations ranging upwards of 5,000 mg/L (5 grams per liter), or about one teaspoon of salt in one quart of water.[45] Freshwater streams become toxic to aquatic animals when watersheds have about 35% of their area covered by impervious surfaces.[46]

Another example of poor stream health comes from an urban area of Melbourne, Australia. The top right graph shows how a measure of stream health depends on important measures of impervious surface. Ecologists monitor water quality and stream health by tallying all of the macroinvertebrates in a stream. They then merge these surveys with species-level scores: A stream scores a 1 for each animal of the few species that can almost survive in a toxic waste dump (the equivalent of aquatic cockroaches) and a 10 for each animal that dies from the slightest whiff of a pollutant. The resulting SIGNAL score indicates the stream's health, with a higher score meaning higher water quality.

The Australian study quantified impervious surface using two different measures. The first type measures total impervious surface covering land in a drainage basin, regardless of where rainwater goes after falling, be it a lawn, ditch, or stormwater drain. The second type, effective or connected impervious surface, is more difficult to measure but also more relevant to water quality because it harms streams the most. This impervious surface fraction drains directly into the stormwater system, and thus directly into streams. With these connected surfaces, stormwater doesn't touch terrestrial soils at all, implying very little chance if any to filter out any impurities. For example, a Durham road's storm drain captures rain that falls and washes particulates and trash and oil from the road, flushing it right into Raleigh's drinking water reservoir.[47] Certainly, some filtration takes place in the streambed, but often the drainage takes place under high rainfall and high streamflow, minimizing even this small opportunity.[48] Even very low amounts of effective impervious surface seriously degraded the Australian streams — when it covers a little less than just 10% of the watershed's area. This amount translates to around 15% for total impervious surface.

Similarly, an extensive survey of streams in Southern California demonstrated an extensive physical transformation of streams classified as urban, those above 8% urbanized watershed, and resulted in the loss of sensitive aquatic insects as shown in the bottom graph.[49]

Durham rainfall exceeds regulated basin sizes.

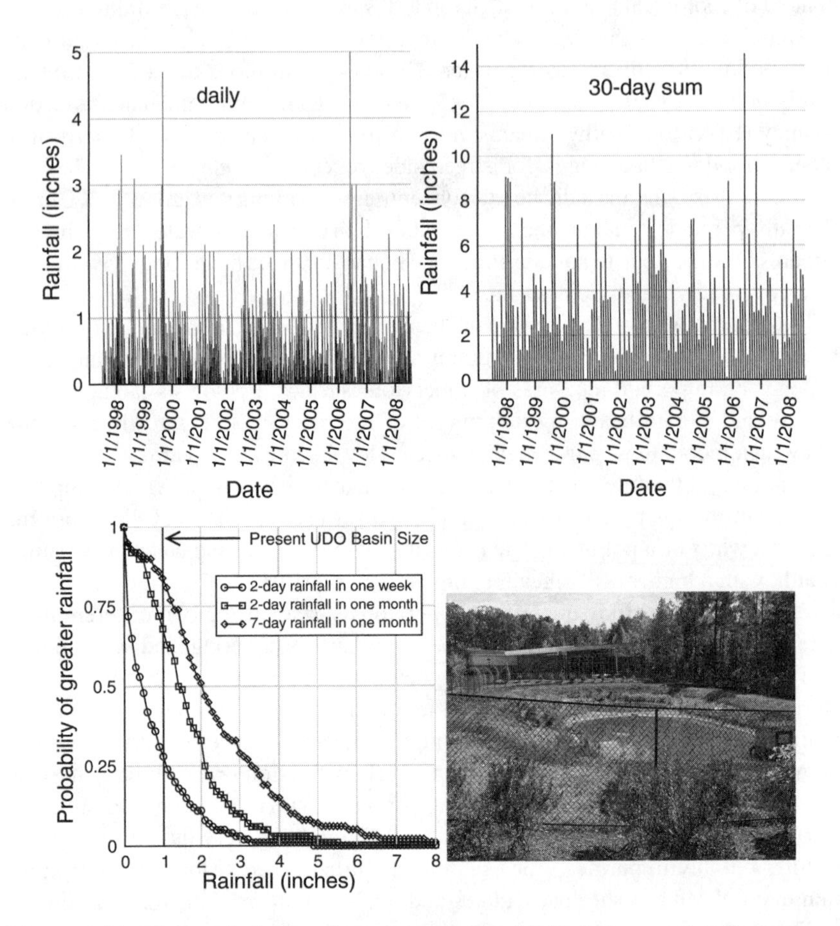

Figure 1.11: The top left plot shows the nearly 4,000 24-hour rainfall measurements from Durham County, and the top right plot adds up rainfall on 30-day periods throughout the dataset. The bottom left plot summarizes the likelihood of exceeding a given amount of rainfall in any several-day period of time within either a week or a month compared to stormwater regulations. The photo shows a retention basin draining a new North Durham library's parking lot: if designed more carefully, an opportune location for frog-chasing, pond-stomping educational opportunities.

Rain washes these pollutants into streams. Durham's yearly rainfall reaches nearly 48 inches, almost 4 inches per month, 1 inch per week, or 1/7 inch per day.[50] Daily rainfall rarely exceeds 4 inches, but often exceeds 1 inch. The top left plot of Figure 1.11 shows the nearly 4,000 24-hour rainfall measurements during this interval from Durham County. The top right plot adds up rainfall on 30-day periods throughout the dataset. Despite the 4-inch average per month, Durham has many months with much less rain, and many months with much, much more rain. It's during these very wet periods, after Durham's clay soils have absorbed all the water they can possibly hold, that additional rains run off into urban streams.

Are Durham's stormwater detention ponds, like that at bottom right, sized properly for Durham's rainfall? The bottom left plot summarizes the likelihood of exceeding a given amount of rainfall in any several-day period of time within either a week or a month.[51] For example, the curve marked by diamonds shows that with 50% chance, rainfall will exceed 2 inches in a 7-day period, and with 25% chance, rainfall will exceed 3.3 inches. During any given week, with 25% chance there will be a 2-day rainfall exceeding 1 inch.

If basins serve to prevent runoff from immediately entering streams, then development rules and ordinances should reflect Durham's risk tolerance for runoff that makes its way downstream to the city's (or other cities') drinking water reservoirs. In other words, supposing a design that removes water only between rainy bouts, Durham's development rule's 3,600 ft^3/acre basin size roughly represents a 1-inch rainfall, a volume exceeded roughly 25% of the time by 2-day rainfalls in any week, and about 84% of the time by 7-day rainfalls during any given month. In contrast, a 7,250 ft^3/acre basin requirement (which holds a 2-inch rainfall) means that, within any week, the basin would be exceeded 10% of the time during a 2-day period, and 50% of the time during a 7-day period in a month. What risk tolerance is appropriate? Should stormwater systems be designed to contain every drop of the 14+ inch rainfall month in 2006, at some unknown cost, or is it sufficient to handle monthly totals of 6 inches?

Citizens must ask similar questions of all types of stormwater systems, including constructed wetlands and infiltration trenches.[52] Whatever option cities choose, trade-offs exist between upfront and maintenance costs, costs borne by developers and costs borne by citizens, and local stormwater costs versus nonlocal downstream pollution and sediment costs (see Figure 1.12).

My photo shows another trade-off: a low-cost, fenced-off, single-purpose retention pond at a public library chosen over a more expensive constructed wetland where trees, birds, frogs, and children may have played.

Reservoirs reduce sediments while providing water.

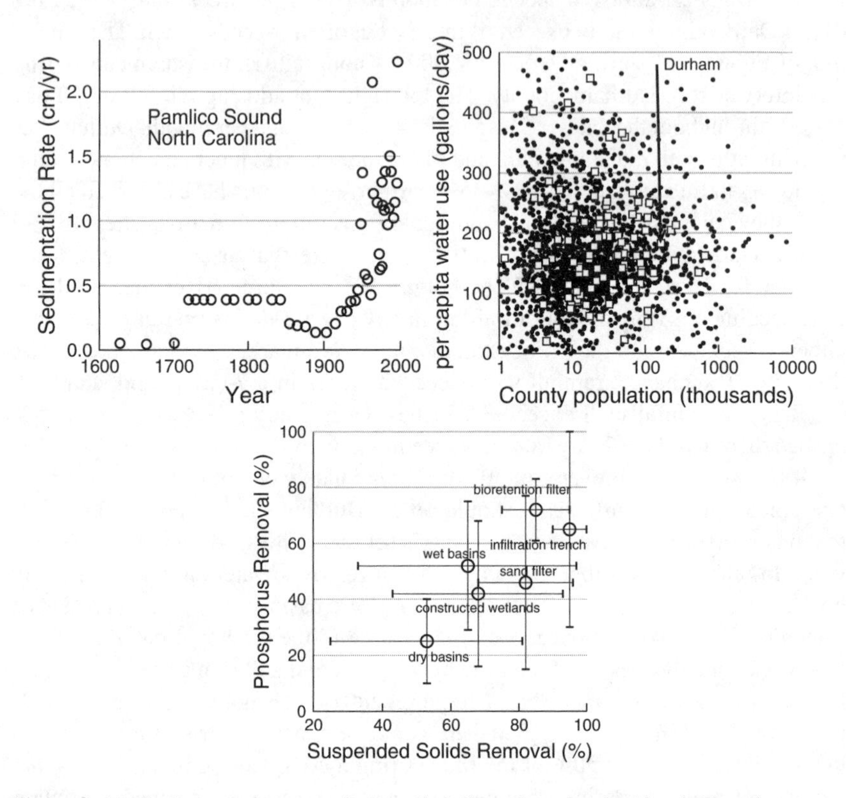

Figure 1.12: Sedimentation rates in Pamlico Sound on the the North Carolina coast increased greatly over the last half century (after Cooper et al. 2004). Durham County's Eno River and Falls Lake reservoir, as part of the Neuse Watershed, ultimately empties into Pamlico Sound. Agriculture, mining, military activities, and urbanization all play a role in this sedimentation, and reservoirs attempt to reduce it. We also drink these waters. Year 2000 per capita public-supplied freshwater use is for 2,384 counties in the United States plotted against its population size (data from water.usgs.gov). Squares indicate the 100 counties in North Carolina. In terms of filtration, things like infiltration trenches (grassy strips surrounding parking lots) and bioretention filters (water-holding gravel filled pits covered with grass and trees) work great (after Weiss et al. 2007).

Stormwater carries pollutants and sediments downstream, and when the water flow decreases, the heavier-than-water sediments fall out, just like the sand being carried in high winds drops to the ground when the air calms. Half of Durham sits in the Neuse River basin, meaning half of the city's stormwater empties out into Pamlico Sound at the North Carolina coast.[53] Sediment cores taken in 1997 (Figure 1.12) show that as much as 2 cm of sediment, nearly an inch, drops into the sound each year! That's at least 10 times more than three centuries ago, with most of that increase taking place in the last half century. Pity the critters living at the sound's bottom.[54]

Across the United States in 2000, the average county had just under 71,000 people with publicly supplied water, and the average such-supplied person in the average county uses about 213 gallons of publicly supplied water per day.[55] Durham County had 167,000 people using public water, with 183 gallons per person per day. At a density of one person per 0.83 acre, 183 gallons spreads out to a 0.2 mm layer on each person's area each and every day. Over one year that equals 7.3 cm — 3 inches — of rain. Durham gets about 48 inches of rain per year, so its citizens take just over 6% of the total rainfall for their personal use.[56] To meet this demand, we fill reservoirs from rain falling on impervious parking lots and tire-abrading freeways. Alternatively, we could preserve rural open space that serves as the watersheds for our drinking water supplies, letting forests and stream buffers filter rainwater. Since we lose about 60% of rainfall to evapotranspiration (Figure 2.14), at least 12% of the county's forested area must be preserved as watersheds. Durham County uses both approaches, preserving rural open space as well as cleaning parking lot runoff.[57]

Constructed wetlands that include trees can serve many purposes: purify drinking water by filtering stormwater, cool the city, support wildlife, as well as provide recreation and environmental education.

The bottom graph of Figure 1.12 compares many types of stormwater treatment approaches.[58] Many competing aspects favor one type of stormwater system or the other. Construction cost data show that bigger is cheaper for all stormwater systems, and annual maintenance costs run about 3 to 5% of construction costs for large systems. One notable exception is that dry extended basins require very low maintenance costs whatever the size. Constructed wetlands are an overall good solution, nicely solving the stormwater filtration problem while promoting wildlife. An integrated approach to a city's stormwater drainage system would be nice, a task complicated by competing goals of different city and county departments. Indeed, if the toxicity levels weren't too high, children could also chase frogs and salamanders in these places.

Chapter 2

Shading and Cooling in City Climates

Most people live in cities containing very high population densities — densities much higher than the natural ecological world expects. What unique environmental properties do cities possess as a result? Here I first demonstrate how much urban areas have sacrificed vegetation — indeed, that's why they're urban — and produced much higher temperatures than surrounding rural areas. In thermal satellite images these so-called urban heat islands (UHIs) poke out of the cooler rural background. Thermally massive urban surfaces, like parking lots and cement buildings, radically change urban environments, making cities somewhere around 5–10C (9–18F) hotter than nearby forests. Interestingly, North American and European cities show distinct urban heat island patterns concerning temperature differentials as a function of a city's human population. Different building styles, measured through a "skyview factor," resolve this difference: At street-level, you can view more sky in European cities, and this difference makes them cooler. In one well-documented example, Los Angeles, California's urban temperature increased about 4C (7F) over the last 120 years. Considering the concomitant population increase, that change corresponds to what one expects from a North American urban heat island.

Urban heat islands can change weather patterns at local, regional, and continental scales, from shifting the time of day that precipitation occurs to inducing localized thunderstorms. Studies from Atlanta, Georgia, for example, even demonstrate that more lightning strikes take place in urbanized areas compared with outlying regions, and lightning strike data suggest a similar pattern for urban areas of Durham County. City atmospheres also influence local climates, just like glass greenhouses and global greenhouse gasses, but particulate emissions complicate the picture by adding a shading effect.

Parking lots, with their extensive impervious surfaces and high thermal mass, represent one location where trees and vegetation can reduce heating and its related issues. However, business owners' profit motives and safety concerns rarely align with optimal shading, and comparing parking lot trees with their natural shading potential clearly demonstrates suboptimal shading.

The heating of impervious surfaces generally matches and confirms the scale of urban heat island warming, and also matches the amount of heat that evaporates a light rain. I then examine the role urban vegetation plays in cooling cities and treating stormwater, and discuss whether any significant energy reduction and carbon sequestration benefits can be expected. Comparing the cooling potential of a tree shows that a tree just can't transpire enough water to cool these high thermal mass surfaces. Further, trees experiencing realistic urban scenarios have an even bigger challenge as their transpiration systems shut down due to high heat. Simply painting these cement and asphalt surfaces white, on the other hand, could greatly change how much heat they absorb. Surprisingly, lawns have a higher potential than trees for cooling via transpiration, and interesting approaches combine parking lots with grass.

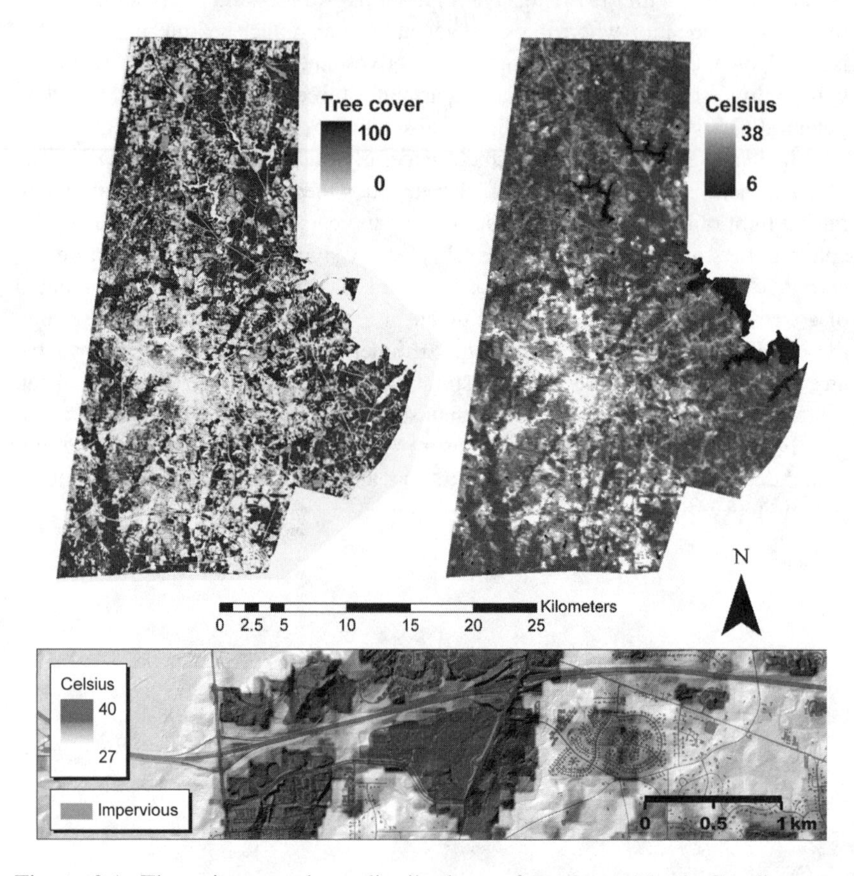

Figure 2.1: These images show distributions of Durham, North Carolina, tree cover in 2005 (30 m square pixels; darker areas depict higher canopy cover) and midmorning temperatures on May 8, 2005 (120 m square pixels; darker areas represent lower temperatures). The warm central core, about 5 km wide, marks the city of Durham. Lakes show up as regions with low canopy cover and low temperatures (see Figure A.2). The bottom image zooms in on Southpoint Mall, with dark areas being warm and small dots indicating impervious surfaces. (All images courtesy of Joe Sexton.)

Impervious surfaces come with urbanization, surfaces that rainwater can't infiltrate, and include anything other than ground cover: buildings, parking lots, streets, sidewalks, bike paths, and garden sheds.[1] With these surfaces come urban heating. and this chapter examines, in part, the resulting "urban heat island."[2] This wonderful example (Figure 2.1) from Durham, North Carolina, demonstrates that high temperature comes with low vegetation, with a nearly 5 km diameter vegetatively depauperated, high-temperature urban core.[3]

Where is this urban heat coming from? I've often wondered whether importing fossil fuels and electrical energy into the city and using it there might contribute to the urban heat island: Think of a toaster with its coils glowing orange in the city, heating up the space within it, but the cord and plug-in leading off somewhere to a coal-fired power plant off in the rural hinterlands. Is human energy use important to the heat island? As the calculations connected with Figure 3.17 demonstrate, an average North Carolinian uses about 83 Btu/m^2 of energy each day. In any event, this is a trivial amount compared with the Sun's energy input of about 13,600 Btu/m^2/day. Human energy input seems pretty minimal, at least for Durham.[4]

This comparison means that the heat in the urban heat island comes from the Sun, and the heat hangs around because impervious surfaces in urban areas reflect and absorb sunlight differently from rural areas. However, not all impervious surfaces are created thermally equal. I have a shed with a clear roof made of polycarbonate plastic: It has no thermal mass to speak of, but it intercepts rain just like my concrete driveway. Rain falling on both surfaces ultimately reaches the ground in my yard rather than going directly into the stormwater system, but the unshaded fraction of concrete also soaks up solar energy during the day and releases it over a long period of time. Asphalt shingles on my house behave somewhere in between these extreme thermal mass examples.

The bottom image zooms in on Southpoint mall in southern Durham County (see the aerial view in Figure 5.7), showing several fascinating points. First, the mall itself, with all its asphalt, is quite warm. Second, neighborhoods see much variation in their heat island fingerprints. Finally, New Hope Creek, which flows into the Jordan Lake reservoir, sits on the left side and shows very cool temperatures. Along this creek bed sit high impervious surface developments, and sharp thermal boundaries are visible between the creek and developments.

Low vegetation correlates with high temperature in Indianapolis.

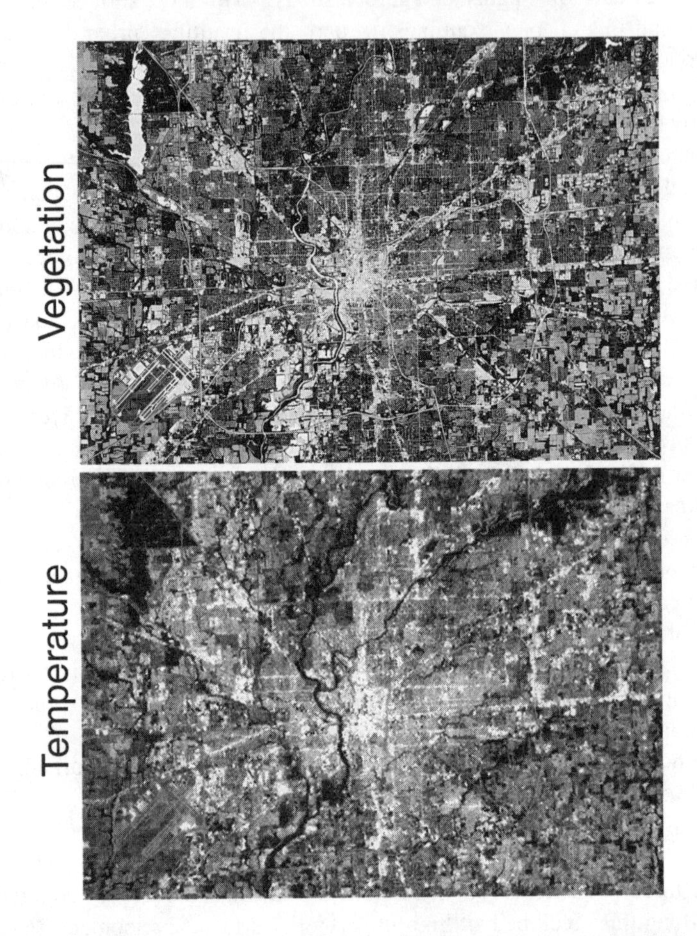

Figure 2.2: Satellite images of Indianapolis, Indiana, taken at 11:15 AM on June 6, 2000, show correlations between vegetation (top image; light areas have low vegetation) and temperature (bottom image; light areas are warm). For spatial scale, note the 3.2 km long lake in the upper left; the total horizontal distance is roughly 18 km. A plot of the numerical data from these two images, in Figure 2.3, indicates the range of temperatures and relative vegetation values. (Images courtesy of Jeff Wilson.)

The city of Durham, North Carolina, isn't alone when it comes to high temperatures. Indianapolis, Indiana, also possesses a correlation between vegetation and temperature.[5] In this case, think of anything but impervious surface as vegetation, including lawns, forests, streams, and lakes, and the lack of vegetation (light areas) in the central portion marks the urban core with residential, commercial, and industrial land uses. Getting vegetative cover and land surface temperature from a satellite sounds complicated. The satellite measures the light (from the entire spectrum) reaching it in its orbit, and this light includes land and atmospheric radiation, the latter needing removal to uncover the portion from land. As for the measure of vegetation, the normalized difference vegetation index (NDVI), and its calculation using GIS software is also rather complicated, involving details of the light spectrum reaching the satellite.[6]

The overall length scale in these images for Indianapolis (Figure 2.2), as well as for Durham in Figure 2.1, is roughly 20 km. That scale brings up an interesting comparison with our atmosphere because the lower 10 km, the troposphere, holds roughly 80% of our air mass. An even shallower layer of the atmosphere, the "planetary boundary layer," defines the upward extent of the atmosphere mixed up by interactions with trees, hills, buildings, and what-not.[7] This layer extends upward only about a few kilometers, quite a bit lower than the height of the troposphere (except, perhaps, in mountainous terrain).

In this image, take note of the unvegetated warm area of 1 km in size right in the center, about a third the length of the long lake at upper left. The urban heat island extends about 1 km high, so project that distance up off the page. Like a deep, slow river flowing over a bed of rocks, the atmospheric flows above the planetary boundary layer don't feel the bumps in the terrestrial surface. These cities' 20-km hotspots are not only wider than the atmosphere is thick, but they're much wider than the normal planetary boundary layer is high. Imagine putting a really powerful hotplate at the bottom of a stream, where it heats water at the bottom that rises toward the surface. A city isn't some small toaster; rather, it's a huge burner whose effects roil the surface, like a huge boulder, with effects felt far downstream. The urban heat island is a virtual mountain of hot air poking out high above the clouds.

In short, anywhere in the world where there's a city, there's an urban heat island.[8]

Low temperature correlates with high vegetation.

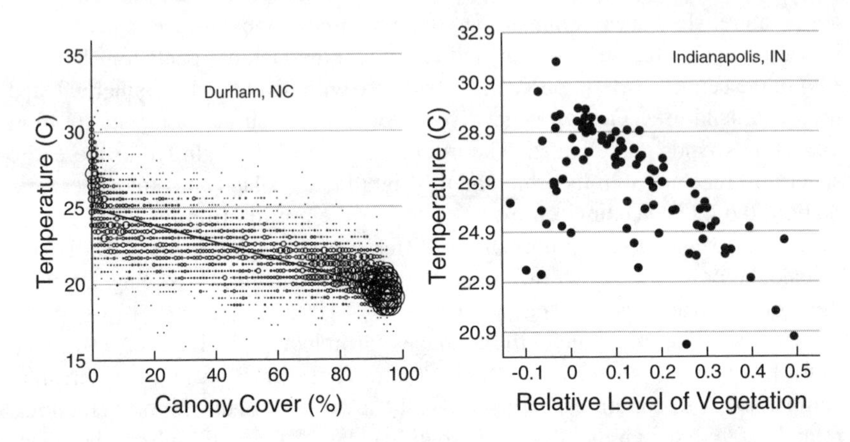

Figure 2.3: These data come from the canopy and temperature images in Figures 2.1 and 2.2: Randomly selected spots in Durham County, North Carolina (top), and various planned developments in Indianapolis, Indiana (bottom). The top plot shows that in Durham, having full canopy cover provides a 5C (9F) temperature reduction compared with having just 10% cover. Removing that last bit of cover, going from 10% to zero, results in several degrees more warming. Indianapolis sees similar effects under a relative measure of vegetation, but with interesting outliers (after Wilson et al. 2003).

The Durham and Indianapolis maps in Figures 2.1 and 2.2 provide nice visual connections between vegetation and temperature, but lack easy interpretations of the correlation strength between vegetation and temperature. Here in Figure 2.3, vegetation and temperature values plotted against one another inform deeper statistical questions. From the Durham, satellite images, sampling just 1% of the pixels provides more than 8,000 points making up the left-hand plot. Many of these points had identical vegetation–temperature values, so I plot larger circles when many locations had identical values.

What do we see? At the middle of the morning on this spring day, the temperature at a location with zero percent canopy fraction was typically 25C, or 77F. In contrast, the temperature in 100% vegetated forest was 19.3C, or 66.7F. Vegetation keeps areas cooler by up to 10F! Though these data come from an early spring morning, I expect similar temperature differences on a hot summer day, meaning a cool 90F in the forest versus a baking 100F in the city.

The Indianapolis plot at right addresses specific urban planning questions. Each of the 88 points plotted here represents a planned development (think "subdivision"), with its temperature and vegetation signature measured by satellite. Outliers in the lower left corner of the Indianapolis plot — little vegetation and low temperature — identify subdivisions with lakes that skew subdivision-averaged temperatures greatly downward. Developments in the lower right have lots of vegetation, keeping neighborhood temperatures cooler, while others have greater impacts on temperature by rather serious decrements in vegetation (upper left region).

Data like these highlight an approach for cities facing urban heating issues as they grow. Local governments could use these data to identify cool features of existing neighborhoods under the conditions unique to their own locations. Armed with this understanding, they could provide incentives to "retrofit" existing hot neighborhoods and demand "cooler" new developments. These neighborhood scale changes could push cities toward the lower right and address problems outlined in the rest of the book.[9]

In the remainder of this chapter, I'll cover the ways cities affect weather patterns, essentially looking at local climate change due to urbanization.

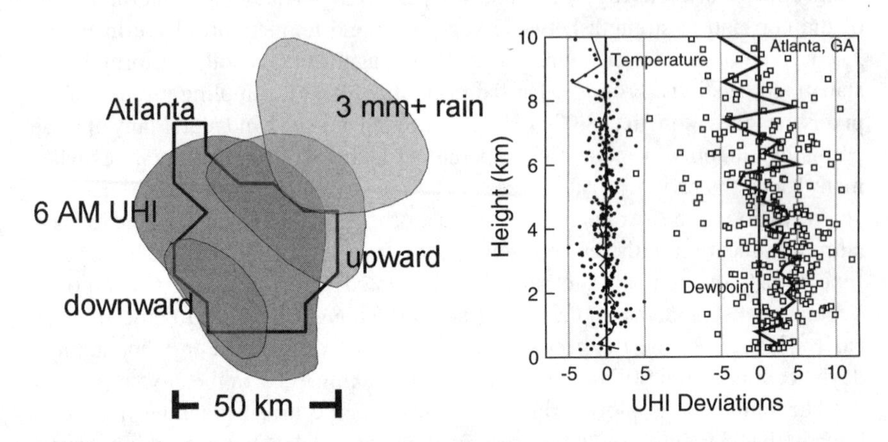

Figure 2.4: The left image depicts an urban heat island over Atlanta, Georgia, on July 30, 1996 at 6:00 AM. Fifteen minutes later, downward and upward air flows developed from the atmospheric instability and generated a thunderstorm lasting from 6:30 to 6:45 AM, with the indicated rainfall pattern (after Bornstein and Lin 2000). The plot at right, measured on 1996–2000 Atlanta summer days deemed ripe for UHI-induced weather, shows atmospheric temperature and dewpoint differences versus height in the atmosphere between thunderstorm and non-thunderstorm days (after Dixon and Mote 2003; data courtesy of Grady Dixon). Data points are fine-scale measures over narrow pressure values; lines join values averaged over larger ranges. Between 1 km and about 5 km in height, days with thunderstorms have slightly warmer temperatures and much higher dewpoints at lower altitude than at higher altitude. This condition reveals the atmospheric instability.

Urban areas affect the weather. The contour plots shown in Figure 2.4, traced from the originals, caricature a thunderstorm that took place during the 1996 Atlanta Summer Olympics.[10] An urban heat island persisted over nearly the entire urban core in the early morning of July 30, around 6 AM, having a central temperature roughly 2.5C to 4C higher than surrounding areas. This heat island produced a thermal instability leading to two regions of high and low pressure, where air moves upward in one spot and downward in another. Think about a rolling boil taking place in a shallow pan of heated water: Heated water in one spot becomes less dense than nearby spots — maybe due to the vagaries of heating or random shapes and pits in the pan's bottom — and as that water rises to the surface, slightly cooler and denser water from the surface sinks to take its place. Similarly, air over the city heated by the heat island processes described earlier becomes less dense in various spots. The warm air mass starts to rise, reducing the pressure at that spot, and the surrounding high-pressure air mass pushes more air into that location of rising air. The rising warm air quickly cools in the upper atmosphere, the water vapor that it holds condenses like the drops on a glass of ice water, and suddenly you've got a thunderstorm downwind of the heat island.

Examining the weather on all the days during May to September 1996–2000, in Atlanta, the graph at right demonstrates this instability. From the entire set, scientists pulled out 569 days as candidates for UHI-induced thunderstorms because of, in part, low-surface air flows. Over the five years, there were 37 thunderstorms — taking place on 20 separate days — confidently attributed to being spawned by the urban heat island. The study specifically excluded other precipitation events, like those coming in with a weather front, and urban thunderstorms occurring simultaneously with rural thunderstorms.

Indeed, the data showed that these thunderstorms occurred on hot, sticky, calm days when the dewpoint[11] was about 3C higher than usual. In these instances, the UHI could be considered high for the high humidity, and the UHI gives the extra push needed to start the convection of the air mass.[12] In other words, just like in a pan of nearly boiling water, atmospheric convection cells spontaneously initiate and generate thunderstorms. Urban heating turns the heat up just a little bit. In this sense, these results suggest that urban weather isn't dominated by UHIs; rather, UHIs slightly shift the thunderstorm-or-not balance.

Cities change rainfall patterns.

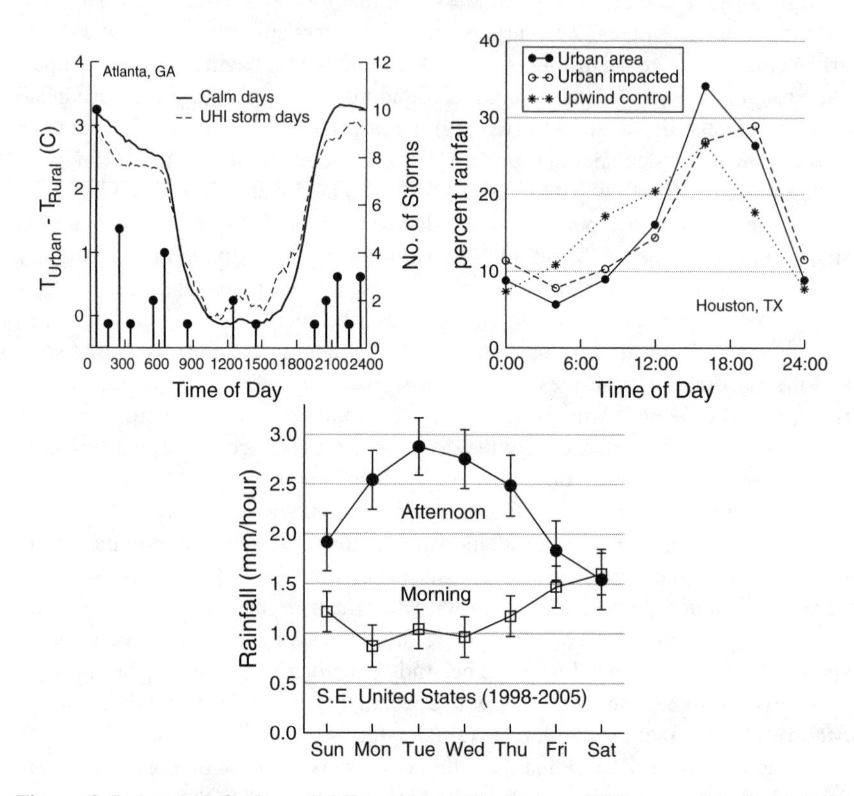

Figure 2.5: At top left, the solid line shows the average UHI over all calm days in Atlanta, and the dashed line isolates days with UHI-induced thunderstorms. Dots indicate thunderstorm frequencies (after Dixon and Mote 2003). UHIs on storm days show rather unremarkable temperature differences. The top right plot shows summer rainfall patterns in and around Houston, Texas, over a 15-year period, comparing an upwind control area (stars and dotted line), the urban center (filled circle and solid line), and a downwind impacted area (open circles and dashed line). Houston has relatively fewer morning showers and more afternoon rains (after Burian and Shepherd 2005). At bottom, overall rainfall patterns from 1998 to 2005 across the southeastern United States show patterns on a weekly and daily scale (after Bell et al. 2008).

The three plots shown in Figure 2.5 provide an overview of fascinating weather changes at local, regional, and continental spatial scales. At top left, the lines compare UHI intensities for calm days and the 24 hours prior to UHI-induced thunderstorm days in Atlanta, Georgia. Sixteen one-hour periods have significantly different temperatures from the nonstorm control days, meaning that not only do UHI storm day profiles differ vertically (see Figure 2.4), but their daily profiles also differ. The urban heat island mostly occurs in the evening, night, and early morning. Urban temperatures intensify in late afternoon, staying high through the early part of the night, with early morning temperatures declining because of radiative heat loss. Furthermore, rural areas cool down rather quickly at night while urban areas retain heat longer.[13]

More evidence for altered weather comes from a study of 15 years' worth of thunderstorms, at top right, around Houston, Texas.[14] Compared with an upwind control region, the urban area and downwind urban impacted region have more afternoon rains, 59% and 30%, respectively, along with 80% more storms between noon and midnight in the warm season. I don't show it here, but comparing these "post-urban" (1984–1999) rainfall patterns with a "pre-urban" period, 1940–1958 also demonstrated that annual precipation decreased by about 10% from 8 PM to 8 AM, and increased by about 14% from noon to 8 PM, a shift most likely attributable to urban heating. Furthermore, between the pre-urban and post-urban periods, summer rainfall amounts decreased by 8% in the upwind control area, but increased by 25% in the urban area. In effect, cities shift around rural rains.

Yet another study, shown at bottom, covered the southern half of the United States, including the Atlantic Ocean and the Gulf of Mexico, using data acquired from the Tropical Rainfall Measuring Mission (TRMM) satellite.[15] Averaging data from 1998 to 2005 demonstrated a strong weekly rainfall pattern in the southeastern United States, along with greatly different morning and afternoon rainfall intensities.[16] In contrast, the southwestern United States displayed no such patterns. It's thought that these cycles arise due to increased air pollution that decreases the rain droplet size by providing many more condensation particles, increasing the length of time a storm develops, and, in the end, increasing the total rainfall. The midweek rainfall reduction, well before the end of the polluting workweek, poses a mysterious question: Complicated land–ocean interactions seem, at present, the best foundation of an hypothesis.[17]

Thus, heat, humidity, and pollution from cities all play a role in changing urban weather.

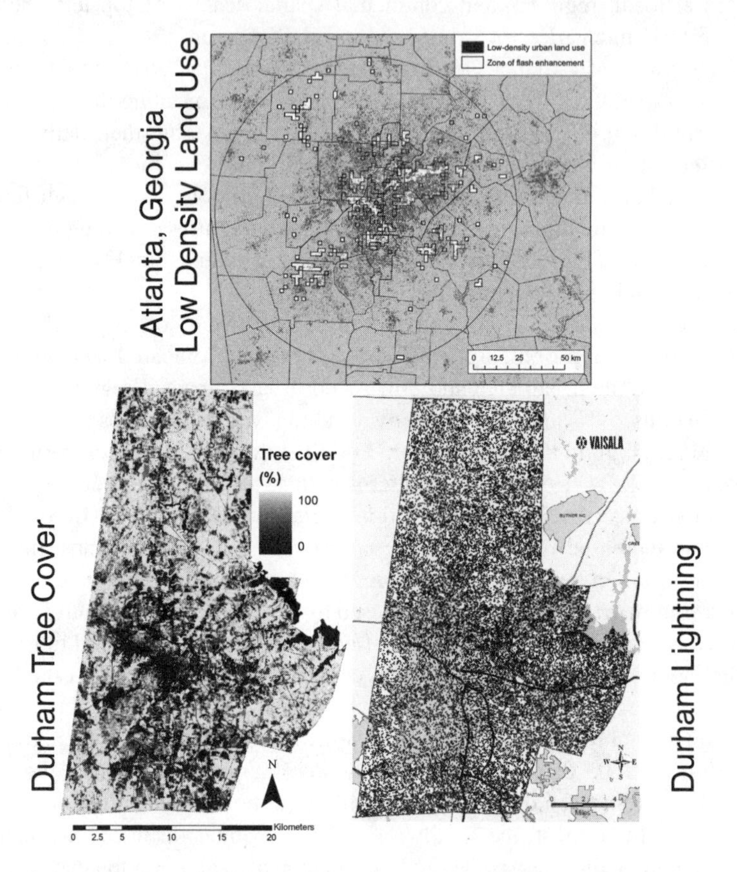

Figure 2.6: The top map shows urban land use and lightning around Atlanta, Georgia. Dark points indicate low- and high–density land use, while each boxed-in white area shows a region of high lightning strike density, identified using the region's 8.2 million cloud-to-ground strikes over the years 1992–2003 (courtesy of Tony Stallins; after Stallins and Bentley 2006). The bottom images show, at left, Durham's 2005 tree cover (courtesy of Joe Sexton) and, at right, locations of the 40,392 lightning strikes observed in Durham County from January 1, 2004, through August 20, 2008 (courtesy of Vaisala, Inc.).

We just examined evidence in Figures 2.4 and 2.5 that urban heat islands spawn thunderstorms that change city weather. Along with thunder, of course, comes lightning, and in another look at how cities affect weather, the top image in Figure 2.6 shows the frequency of lightning strikes in Atlanta over a 12-year time span.[18] During these 12 years, automated equipment measured around 8.2 *million* cloud-to-ground lightning strikes, with an estimated 50 to 75% more lightning strikes in urban land-use areas![19] This figure plots, along with the strikes, the visual correlation with one measure of urban areas, low-density urban land use.

Mechanistically, urban land use has lots of impervious surfaces that generate urban heat islands that spawn thunderstorms and produce lightning strikes. These strikes occur, therefore, right where people live.

The correlation turns out to be a bit more nuanced than just the existence of impervious surfaces. Much of the housing development in Atlanta has taken place downwind of the major city center that generates the UHI; that situation reinforces the correlation between where people live and increased lightning, but does not really affect the mechanism behind the correlation.

Although there are no statistics, the bottom images show 2005 canopy cover (dark areas mean fewer trees and more urbanization) and the 40,392 lightning strikes in Durham County over a four-and-a-half year period.[20] Do the strikes occur in the urban areas where there's no canopy? If I squint, I see a downwind correlation, but here's why science needs statistics: Is there really a correlation, and might something like topography be more important?

Given that urban areas have more lightning, let's put this it into an economic context. The damage costs associated with lightning strikes in the entire state of Georgia, not just Atlanta, is estimated to be about $18 million per year, according to insurance claims, and around $332 million annually across the entire United States.[21] Though this $332 million is a large number in aggregate, it works out to just over $1 per person in the United States. The Atlanta metropolitan area's population is around 5 million, covering just under 22,000 km^2, meaning Atlanta's lightning imposes costs around $5 million per year. Lightning strikes might be worthwhile to consider while building cities, though it seems inherently difficult to plan housing developments around their lightning production and attraction characteristics. In the end, if these cost estimates are accurate, the economic burden doesn't seem inordinate — something like $200 per square kilometer — unless it were my house that was hit by lightning.

Cities grow warmer.

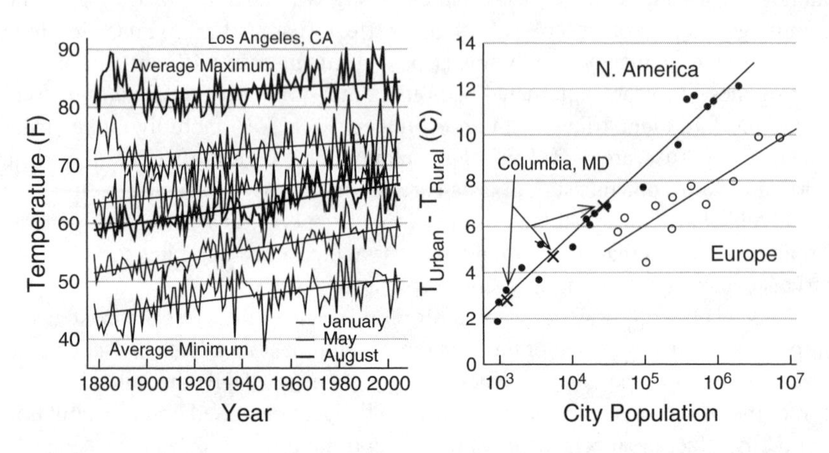

Figure 2.7: At left are historical temperature extremes measured in Los Angeles, California (after Akbari et al. 2001). The six lines show average minimum temperatures (bottom three lines) and maximum temperatures (top three lines) for the months of January (the thin lines), May (the thicker lines), and August (the thickest lines), for the years 1877 to 2005. Temperatures increased over the long term, but the minimum temperatures increased twice as fast. At right, the temperature difference between hot cities and cooler, nearby rural areas defines the urban heat island (UHI). Plotting this difference against the logarithm of city population reveals two straight lines, one for Europe and one for North America (after Oke 1973, 1982). Three data points for Columbia, Maryland (denoted by Xs), plot changes taking place with its population growth over the years. Its trajectory closely follows the North American relation between city size and urban heat island.

Over the last few pages we've seen the connections between land-use type, vegetation, and urban climate. Data considered in Figure 2.7 involve citywide averages over much longer time spans, and we find that as cities grow, so do their urban heat islands.

Over the last century, minimum and maximum temperatures increased in Los Angeles, California,[22] where minimum temperatures rose twice as fast as maximum temperatures during the 130-year span shown in the left plot. The straight lines plot statistical fits to the data, giving fitted slopes of minimum temperatures as 0.48, 0.63, and 0.63 F/decade, and maximum temperatures as 0.38, 0.26, and 0.21 F/decade.[23] There are puzzling features, such as the decreasing maximum temperatures in May and August up until about 1910, but, overall, temperatures increased over the last century by about 5–10F.

At least two possible reasons might explain this increase — global warming (discussed in Chapter 3) or the just-discussed urban heat island. Let's first consider the possibility of local warming due to the urban heat island. As long as 25 years ago, scientists knew that larger cities experienced larger urban heat islands,[24] just from looking at the temperature difference between urban and rural weather stations. European cities seemingly defy the laws of thermodynamics, at least as practiced in the United States, but I later show data resolving that peculiarity (see Figure 2.8). The North American data plots one city, Columbia, Maryland, at different population sizes as it grew over the years. With its population, so, too, grew its heat island.[25]

Of course, Los Angeles has changed dramatically over this period. The U.S. Census reports L.A. County populations for 1900, 1950, and 1995, as 170,000, 4.2 million, and 9.3 million, respectively, or about a fifty-fold population increase.

Does this population growth account for the 4C temperature change seen over the last century? Given L.A.'s fifty-fold population increase, the urban heat island regression suggests an expected temperature change of roughly 6C for U.S. cities and 4C for European cities. These numbers fall right within L.A.'s observed temperature increase, so, it seems, we can't easily dismiss the idea that urban heating explains the increased temperatures of Los Angeles over the last century.

Closed-in urban areas have higher heat islands.

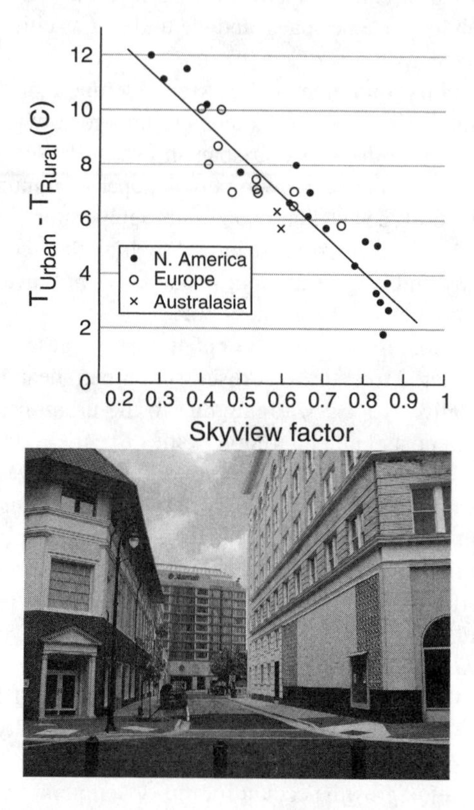

Figure 2.8: My photograph of a closed-in street in downtown Durham depicts a small skyview factor — the small patch of viewable sky versus all the walls blocking a larger skyview. This skyview factor (0 means no viewable sky, and 1 means a fully open sky) resolves the North American–European difference seen in Figure 2.7. Plotting a city's urban heat island effect against its skyview factor, instead of its population, results in a single curve for all three regions plotted here (after Oke 1982). Evidently, large European cities possess shorter buildings with more open-sky views.

The intensity of the urban heat island with city population shows marked differences between North American and European urban heat islands as depicted in Figure 2.7. No, America hasn't suspended the laws of thermodynamics. Rather, the urban heat island depends on the "closed-in-ness" of a city, called a city's skyview factor, the fraction of the upward hemispere open to the sky. Imagine the difference in skyviews between an open prairie and a closed-canopy forest: The skyview factor measures the urban equivalent when viewing the sky from the street, with marked differences between residential suburbs and downtown business cores with tall buildings. As an example, I took the photograph shown in Figure 2.8 in one of the few places in downtown Durham with a smallish skyview factor. Standing between these two buildings, walls fill much of the overhead view, giving a relatively small skyview factor.

Skyview factors resolve the disparity between the UHIs for European and North American cities. Indeed, perhaps Los Angeles's urban heat island, discussed in Figure 2.7, might be more in line with Europe's, as my calculation on p. 43 suggested, because it has relatively shorter buildings due to earthquake dangers.

These closed-in city streets are called urban canyons, complete with their own interesting properties related to light and energy. First, I present a short primer on the light coming from the Sun. Just like around a campfire where kids feel the heat of the flames, the Sun's hot surface bathes Earth with 6,000C light,[26] composed of ultraviolet, visible, and infrared wavelengths. All frequencies of light have the same speed,[27] and the product of frequency and wavelength equals the speed of light. High frequency goes with short wavelength (ultraviolet), and low frequency goes with long wavelength (infrared). Photons from sunlight have relatively high frequency and short wavelength. When concrete or asphalt absorbs this high-energy light, the material heats up and reradiates low-energy, infrared light characteristic of outdoor temperatures around 25C. This light is the same thing as visible sunlight, but has a lower frequency and longer wavelength, and thus lower energy and is not visible.

Out in a parking lot this infrared light radiates from the asphalt in a hemispherical way, up into the sky, but in an urban canyon the light radiated from the wall of a building often moves toward the wall of a neighboring building or toward the road or sidewalk, and gets reabsorbed. This process of absorption and emission goes on and on. Light bouncing around urban canyons delays the cooling of the roads and buildings in a city core. Cities, of course, can solve this problem with strategically placed vegetation (see Figure 2.17).

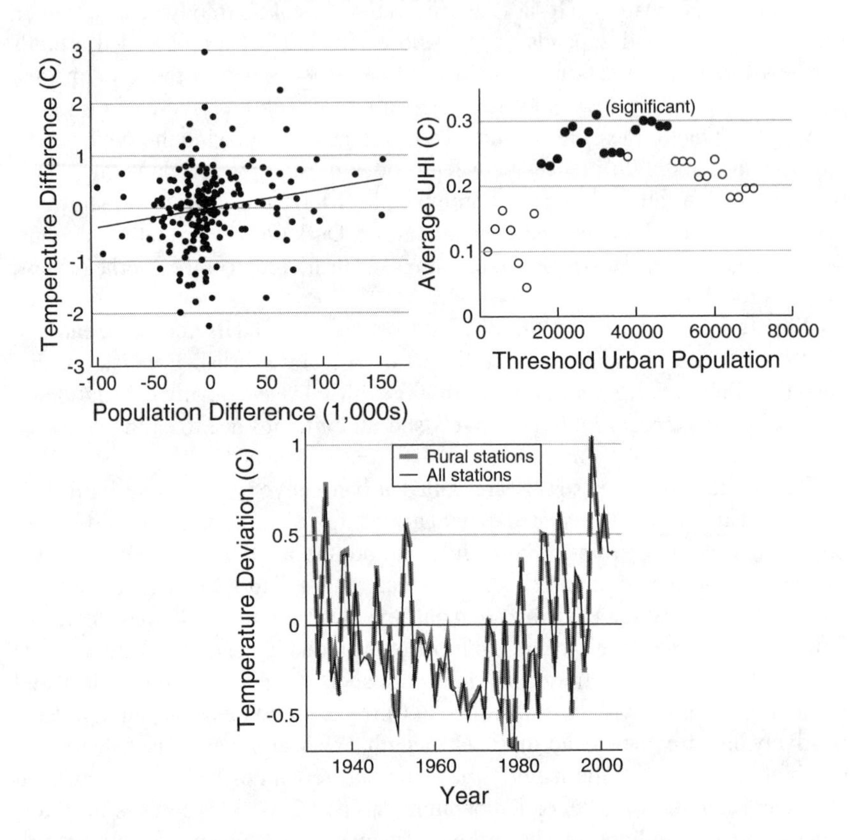

Figure 2.9: Using clusters of weather stations in the top left plot, a station's temperature and population difference from its cluster average shows a weak correlation. Another analysis of these clusters, in the top right plot, shows a small urban heat island for stations with at least 20,000 people (filled circles). These results demonstrate, using weather station measurements, that an urban heat island exists. Regarding global temperatures, the bottom plot separates urban and rural stations at 30,000 people within 6 km in the year 2000. The curves show little difference (dark sits on top of gray), meaning the UHI didn't cause the measured temperature increase over the last half century (after Peterson and Owen 2005).

Despite the seemingly clear results of Figure 2.7, the urban heat island's existence has been questioned[28] based on arguments that the UHI arises from measurement biases. These biases aren't scientific misdeeds; rather, complications abound in measuring temperature. Once these many problems are corrected, the argument goes, there's no difference between urban and rural areas.

While we consider these monitoring station studies, temporarily disregard the satellite images showing urban heat islands.[29] Clusters of measurement stations are scattered all around the United States. For each station of each cluster, the number of people living within 6 km can be counted, and its temperature and population difference from its cluster's average calculated. The top left curve in Figure 2.9 plots these pairs of differences. An urban heat island exists, however weak, with at most 0.5C temperature differences. Another approach defines a threshold number of people living within 6 km of a weather station differentiating urban and rural stations. Using this threshold, data support the existence of an UHI for populations of 14,000 to 50,000 people in a circle of 6 km radius, but again, the heat island effect was small relative to Figure 2.7 UHI values.

Other studies compared UHIs in different places. In one such example, an examination found strong UHIs in the eastern United States but not in the west.[30] Similarly, coastal differences could cause a nationwide study to wash out important effects in one area due to no effect in other areas.

Let's reconsider the Los Angeles temperature rise, which is seemingly consistent with the growing UHI of a growing population. Do those results mean that, perhaps, there is no global warming, and the observed global temperature rise might just reflect the effects of the urban heat island, an effect that has gotten worse because our hugely increased population (see Figure 1.1) resides mostly in cities? Perhaps easily fooled climate-change scientists have simply been sloppy with their science.[31] Using this collection of weather stations with differing populations within 6 km, the bottom plot compares two average temperatures over the last 70 years: that of all stations, and only those "rural" stations, having populations less than 30,000 people in the year 2000.[32]

About 84% of the weather stations were thus classified as rural. Essentially no difference was found between the two averages, but there is a temperature rise of about 1C beginning in about 1970 in both averages.

The nationwide (and presumably global) temperature rise over the last few decades, clearly present in Figure 2.7, isn't because of measuring increasing city temperatures as cities grew and thinking that increase was a global phenomenon. Increasing global temperatures *are* a global phenomenon.

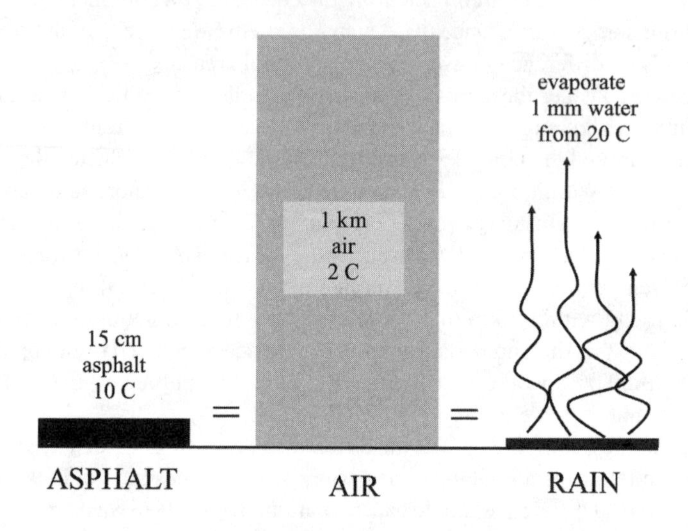

Medium	mass=density×volume	specific heat	temperature increase
Asphalt	$(2000\ kg/m^3)(0.15m \times 1m^2)$	0.8 Btu/kg/C	10 C
Air	$(1.2\ kg/m^3)(1000m \times 1m^2)$	0.95 Btu/kg/C	2 C
Rain:heat	$(1000\ kg/m^3)(0.001m \times 1m^2)$	4.0 Btu/kg/C	80 C
Rain:vaporize		2100 Btu/kg	heat of ⇐vaporization

Figure 2.10: Temperature changes involving equivalent heat contents in asphalt, air, and water. Suppose, in keeping with Figure 2.11, a 6-inch (0.15 m) thick square meter of parking lot asphalt heats up 10C during the day. If that heat were transferred uniformly to a 1 km tall air column above the parking lot, a 2C temperature change would result. After the parking lot heated up, a 1 mm, 20C rainfall could absorb all of that heat if it completely evaporated. The table provides all the numbers needed to do the math. Can transpiration of water by trees provide enough cooling to replace unreliable rainfalls?

Talk of urban heating and contemplating options for cooling means we need a basic discussion of temperature and heat energy. Though technical, a calculation of equivalent heat in each medium reveals important urban realities. Suppose we have a 6-inch (15 cm) thick road or sidewalk surface that heats up by some amount, say, 10C (see Figure 2.11). Further suppose that heat can go one of two places: either to increase the urban air temperature, or to heat up and evaporate water from an afternoon summer shower. Let's compare the numbers.

The heat needed to change an object's temperature depends on the product of three terms: its mass (1 gram, say, versus 1 kilogram), its "specific heat," which roughly means its ability to hold heat (consider wood [low] versus concrete [high]), and, of course, the temperature change (1C versus 10C). Mathematically, we write this product for the heat involved in a temperature change ΔT ("delta tee") as $mc\Delta T$, where the material considered has mass m and specific heat c.[33]

The table in Figure 2.10 summarizes the inputs into my calculations, and I provide them here for easy reference. The first three lines, asphalt, air, and rain:heat, compute heat content from raising temperatures. Multiplying density by volume — thickness times a meter-squared area — yields mass. Mass times specific heat times temperature change for each medium gives the required heat. In addition to heating rain to the boiling point, turning the water into vapor — the steam coming off a road during a hot day's sprinkle of rain — requires more heat, noted as rain: vaporize. Multiplying water's "heat of vaporization" by the mass of rain, then adding the result from the third row, gives the total heat needed to warm and evaporate a sprinkling of rain.

In the end, there's roughly equal heat content in a 6-inch slab of concrete raised 10C, a kilometer high air column raised 2C (see Figure 2.7), and a 1 millimeter rainfall evaporated from 20C: 2,400 Btus vs. 2,280 Btus vs. 2,420 Btus.[34]

Notice that turning water at the boiling point into steam takes a lot more heat than getting the water to boil in the first place. Think about your personal experiences of heating water to the boiling point, taking relatively little time, but boiling all the water away takes much longer. That's the difference between heating water and vaporizing water (read: turning it into steam). It's my guess that during the summer thunderstorms dousing hot roadways, the first little bit of rain that falls quickly cools the pavement, in part, by evaporation, and the remaining rainfall has little cooling left to do, bringing temperatures down.

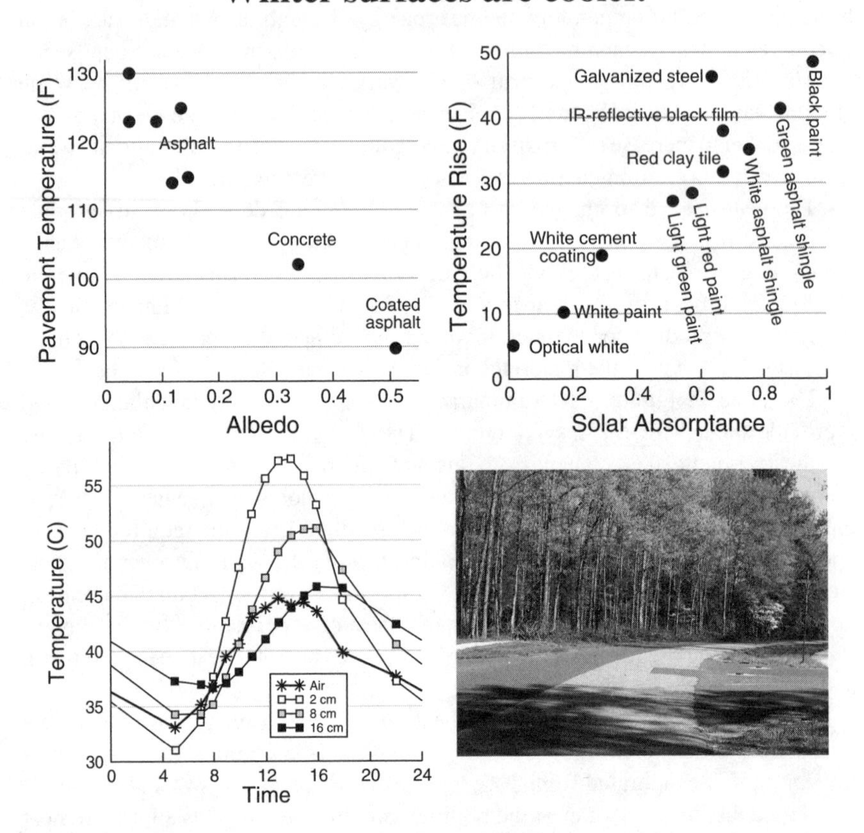

Figure 2.11: Pavement temperature decreases with increasing albedo, a zero-to-one measure of whiteness and reflectance (upper left), and, equivalently, dark objects heat up more because they reflect less light (upper right; solar absorptance equals one minus its albedo) (after Pomerantz et al. 1999). Temperature profiles in an asphalt road show that the upper 16 cm contains much heat, also demonstrating the time lag associated with the urban heat island (after Al-Abdul Wahhab and Balghunaim 1994). My photograph of a road in Durham, North Carolina, shows evidence that asphalt bleaches with age, presumably with different temperatures.

Cities heat up because thermally massive impervious surfaces absorb sunlight. At that surface scale, measurements taken one midafternoon in mid-September in Berkeley, California, shows, in the upper left plot of Figure 2.11, that painting things white reduces the material's temperature (also see Figure 3.14). We all know dark things heat up more; nonetheless, data is data, and the data show in numbers how high absorbed sunlight increases temperature, here as much as 40F.[35]

I took this photo in suburban Durham to show how asphalt naturally bleaches, changing its albedo. Here, the two sides of the road have very different albedos, I suppose, because of different paving dates. Of course, the asphalt mixtures could be colored differently, but I doubt it in this case. When asphalt ages it whitens, and according to the upper left plot, the lighter surface has a lower temperature, reducing overall urban temperatures. Sometimes this bleaching takes place naturally,[36] as shown in the road photo, and other times it needs active management. Scaling this whitening phenomenon up to the city level means that painting things white should reduce urban temperatures.

Consider these measured temperature rises within the road itself. Thick impervious surfaces hold heat down to a depth of about 15 cm, or 6 inches, according to one profile measured in the Saudi Arabian desert.[37] In addition to the depth of the thermal profile, one sees how the lower depths experience a delay in heating up, toward late afternoon. As a result, all of that heat gets stored in this thick thermal mass, being slowly released throughout the evening and nighttime (Figure 2.5).

Later on I discuss more cooling possibilities of both trees and white paint, and show examples of the energy savings that can result. Though it might be cooler, living in a white-painted asphalt and concrete environment might be unpleasant visually, requiring a good pair of polarized sunglasses. White paint certainly isn't a perfect cooling strategy. I show an overhead view of Durham's Southpoint Mall with its light-colored roofing in Figure 5.7, but its thermal image still shows very high temperatures in Figure 2.1. Appropriately placed shade trees might provide the same benefit as white paint, preventing surfaces with high thermal mass from absorbing the of sunlight's energy. Tree leaves also reflect sunlight, turning back as much as 40% of the incoming light and absorbing much of the rest.

Parking lot trees could provide shade.

Figure 2.12: Two parking lots in Durham, North Carolina: The top photograph shows trees in a small mall's parking lot. Notice that the pruning allows shade to fall primarily on the mulched surface below the tree, and the tree's height must be limited to not interfere with the lights above. The bottom photograph shows Northgate Mall's large parking lot with absolutely no trees. In both cases, nearly every drop of rain flows directly into stormwater systems, washing pollutants, oil, and what-not with it, and the unshaded high thermal mass surfaces contribute to the urban heat island effect.

Parking lots cover about 10% of land area in cities,[38] and these unshaded impervious surfaces help create urban heat islands. The streets at Southpoint Mall (and its parking lot) in Figure 5.7 show up bright in the Durham County thermal image (see Figure 2.1). Durham's newer parking lots seemingly have more potential for tree shade than older ones, yet as the top photo in Figure 2.12 shows, parking lot trees still involve several conflicting issues. First, trees aren't pruned for optimal shade. Second, if they were, then, in this case, they would block the above lighting. Resolution of the conflict between reducing daytime lighting and increasing nighttime lighting apparently tilts toward nighttime lighting. Third, the mulched surface surrounding the trees receive most of the shade, precisely where the shade's not really needed. Fourth, that mulched surface could absorb parking lot rainwater runoff, but the curb prevents infiltration. Finally, notice how many parking spaces in these photos go unused (see also Figure 5.7). Perhaps the developers have high hopes when they build these lots, but unused spaces cause many problems that affect everyone.

Durham is not alone in this respect. In 1983, Sacramento, California, passed an ordinance stating that 15 years after issuing a development permit, parking lots must shade 50% of their paved areas. How well did it do? Nearly 20 years later, the first results emerged from a detailed examination of 15 parking lots located in a variety of land-use areas. These lots had a combined 28.7 hectares of paved area, 6% more parking than required by ordinances. Of this, the shading ordinance required 14.4 hectares of shade, but shading covered only 4.1 hectares — just 14.4% of the paved area. Granted, the results included parking lots newer than 15 years old with young trees, yet projections of future shade provided an estimate of just 7.8 hectares of shade, or 27% of the paved surface. Furthermore, covered residential parking spaces provided 44% of the total shade, meaning that protecting a vehicle's shine, rather than reducing the urban heat island, motivated a large part of the shade. There's a clear difference between passing Planning Department regulations and subsequent implementation and enforcement of those regulations.

Regarding costs and benefits, this Sacramento study calculated annual benefits of $19.20 per tree, and if Sacramento met the required 50% shading level, city-wide annual benefits would total $4 million. Yet, even the existing trees in these parking lots needed maintenance. Within the 15 lots examined, 42 trees needed removal, 435 staking adjustments, 41 trimming, and 620 other types of care. Projecting the costs of these unmet needs to the entire city led to a cost of $1.1 million for the trees providing 9% cover. A cost-benefit analysis demonstrated an estimated $20 million tree-planting cost to achieve 50% cover, with a payback time of 10 years.

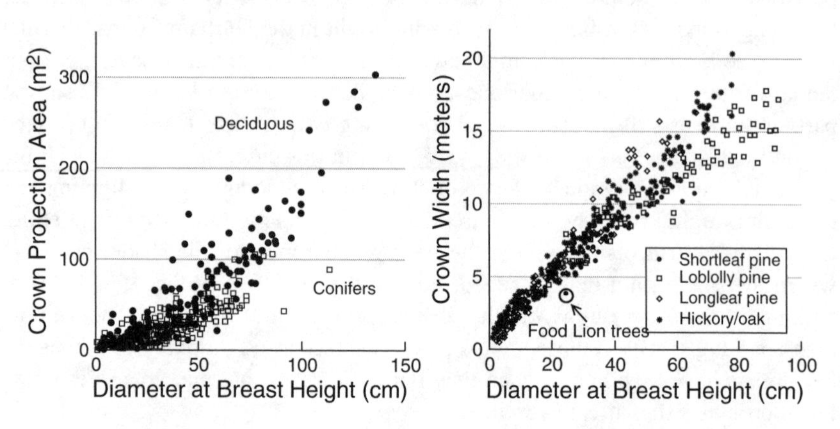

Figure 2.13: The left graph shows that big, deciduous trees (filled circles) shade a lot of the ground below them (after Shimano 1997). The crown projection area, simply the shaded area beneath a tree at high noon, beneath conifers (open squares) never gets as large. The right plot shows another study of how the width of an open-grown tree's crown (diameter) varies with the width of its trunk (data from Smith et al. 1992 and Krajicek et al. 1961). Trees from the Food Lion parking lot (Figure 2.12) are not living up to their potential.

How much space can a shade tree shade? These basic physical measurements on trees indicate trees' natural potential for shading parking lots. The left graph in Figure 2.13 come from a 1966 study of temperate forests in central Japan. The deciduous tree data are from 247 individuals of six species including a beech, *Fagus crenata*, and an oak, *Quercus crispula*, and the conifer data come from 596 individuals from six species. At the high end are some really big trees — 4-foot diameter — much, much bigger than the ones in my photo of the parking lot of a Durham grocery store (Figure 2.12).[39]

The graph at right shows data for trees grown alone, with no competition between trees for sunlight, meaning they can grow as large as possible.[40] Trees experiencing such competition grow taller and narrower, providing less shade, but competition is the least of a parking lot tree's problems. Assuming that trees are trees everywhere, I've combined data on oaks from a study in Austria and data for several pine species that are important for southeastern U.S. forests.[41] This figure includes trees closer in size to my parking lot trees.

Certainly, if one hopes to shade parking lots, one might not choose a loblolly pine. Old trees of this species have the shape of a big tall stick with a cotton ball at top, at least the ones remaining after all their neighbors get chopped down. Such a shape provides very little shade. Furthermore, loblolly pines drop their needles during summer droughts, and big old branches drop from these tall heights, which would do a nasty job on someone's expensive sports car. The resulting lawsuit alone, from the business owner's perspective, would wipe out any environmental benefit. In any event, even though loblolly pines grow fast and survive well in their native southeastern U.S. environment, other native tree species provide much more shade.

Do the Food Lion parking lot trees live up to their potential? I measured the first four trees (Bradford flowering pears) by the shopping cart in that photo,[42] obtaining an average diameter at breast height (DBH) of 24.5 cm and a crown width of 3.7 m, which gives a calculated projection area around 11.5 m^2. Granted, my DBH measures weren't quite at breast height because they branched at a lower height, but their averaged datum point shows they provide relatively little shade for their size.[43]

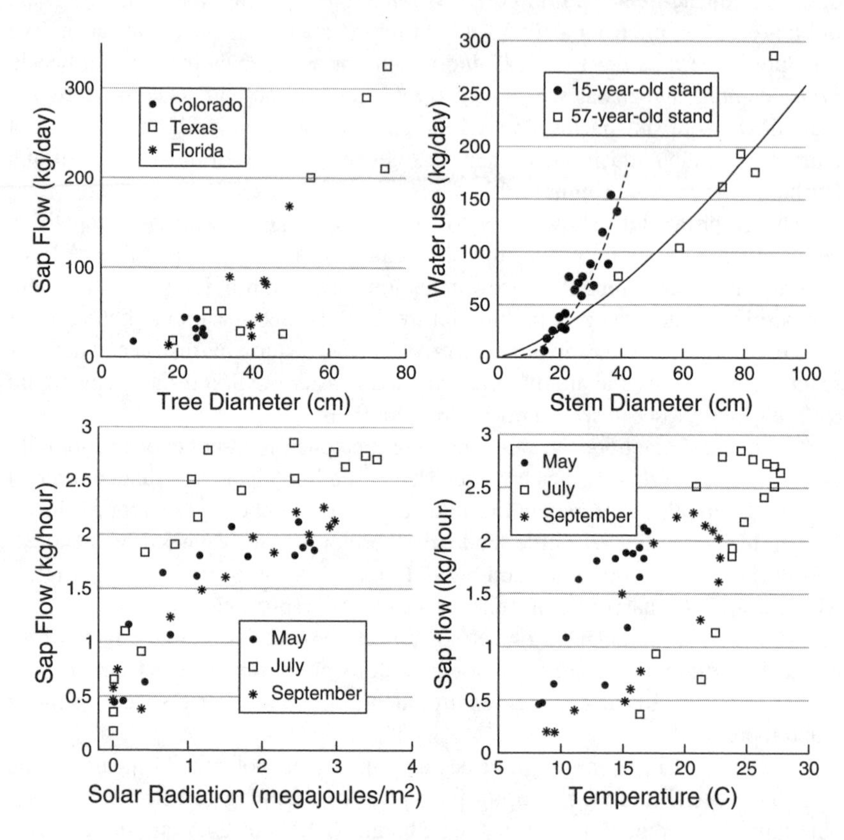

Figure 2.14: The top left graph shows the amount of sap flow in midsummer versus tree size, measured as the diameter at breast height (DBH), averaged over several species from several places (after Vose et al. 2003). At top right, a comparison of two *Eucalyptus regnans* stands shows that water use decreases with stand age (after Wullschleger et al. 1998; stem diameter resembles DBH). The bottom plots show sap flow, from several tree species in central Colorado, as a function of light (bottom left) and heat (bottom right) during spring, summer, and fall.

Now let's come back to cooling. Figure 2.10 showed how evaporation, the all-important cooling method for our own bodies, takes up a lot of heat and in the process cools whatever object gives up the heat. How much potential exists for trees to cool cities through their transpiration? Recall that transpiration involves water evaporating from the small pores in leaves called stomata. Evaporation of water out these pores draws sap up the tree, with the soon-to-be-evaporated water entering the roots and coming from the soil (Figure 3.5). The graph at top left in Figure 2.14 shows sap flow rates for varying tree sizes, averaged over a dozen or so tree species from several locations across the United States. Data in the top right graph show that young Eucalyptus trees transpire more than old ones, thus, perhaps, providing better cooling services.[44] For scale, the parking lot trees I showed in Figure 2.12 have a diameter of about 25 cm.

As a tree's environment changes, so do its photosynthesis rates and, thus, its transpiration rates. The bottom graphs show that more light and heat cause trees to transpire more water.[45] Solar radiation maxes out at full sunlight, but trees start to fail and die at too high of temperatures, but beyond the range shown here (see Figure 2.15).

According to these graphs, we might hope for a sap flow of about 30 kg per day from a 25 cm DBH tree, and, from Figure 2.13, we might hope for a natural crown width of about 6 m, meaning a crown projection area of about 30 m^2. Dividing that water use by the crown projection area gives roughly 1 kg/m^2/day, working out to about 1 mm of water per m^2 each day. A bigger tree of 60 cm DBH also provides about 1 kg/m^2/day transpiration.[46] At either size, the sap flow matches the 1 mm of evaporating water needed to cool a paved surface, as I calculated in Figure 2.10. Trees cool!

Of course, this calculation ignores the fact that if the tree were *shading* the asphalt, then the asphalt wouldn't need cooling, but the motivating hope was that trees provide significant, citywide cooling benefits.

Here's yet another hopeful approach: Reduce the urban heat island (in the absence of rain or shade) by pouring water on hot roadways (assuming water sources aren't constrained). A quick estimate shows that a 4,000-gallon water truck would cover about 2 km (1.2 miles) of an 8 m wide road with 1 mm of water. Perhaps in high-impact areas such a strategy would be worthwhile, but it seems unlikely, given what would certainly be an inefficient endeavor in terms of labor, transportation costs, and added vehicle emissions in hot urban areas.

Trees near asphalt stop transpiring early in the day.

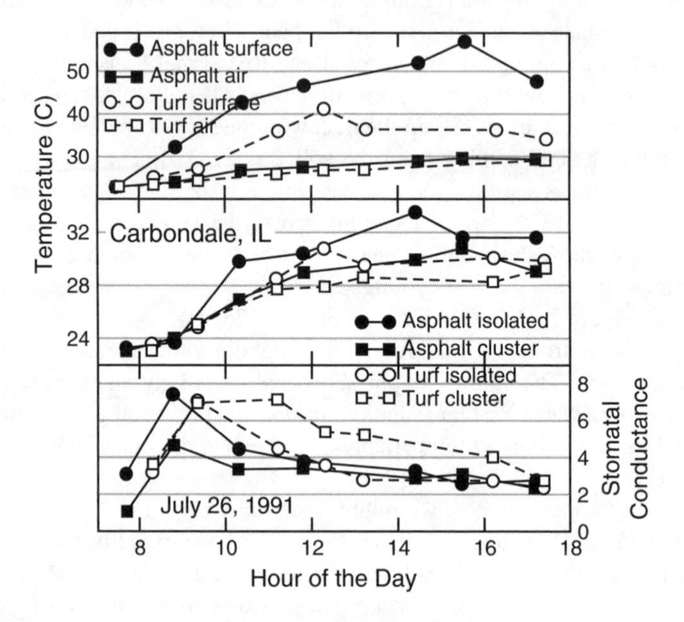

Figure 2.15: Temperature and stomatal conductance of trees surrounded by turf or asphalt on July 26, 1991, in Carbondale, Illinois (after Kjelgren and Montague 1998). The top graph shows the surface (circles) and air temperatures (squares) measured over a 60 m $\times$ 110 m paved parking lot (filled symbols) and a nonirrigated turf lawn (open symbols). "Stomatal conductance" measures, in essence, the water flowing out of a leaf's stoma.

Trees are living organisms with traits shaped through natural selection. Evapotranspiration rates shown in Figure 2.14 came from happy forest trees living in the right environment. Unless very carefully chosen and maintained, planting a tree in a parking lot places it far from these ideal, evolutionarily appropriate, environmental conditions. How does a tree deal with urban environments?

The top graph in Figure 2.15 shows surface and air temperatures at a paved parking lot and an unirrigated open-turf lawn with normal sunlight levels.[47] At the surface, parking lot temperatures (measured by an infrared thermometer) exceed the turf's by as much as 25C, yet for air temperatures at 2 m height (6 ft), air movement equalizes both temperatures to broader atmospheric conditions. Also note the delayed decrease in the asphalt surface temperature relative to the turf surface temperature — another demonstration of the thermal mass effect.

Now put trees into the picture. Researchers placed potted flowering pear trees, *Pyrus callyerana*, on the two surfaces either isolated from other trees or clustered within a group of eight other trees. The middle plot shows that trees grouped over turf, perhaps not surprisingly, experienced the lowest leaf temperatures, whereas an isolated tree over asphalt saw the highest leaf temperatures. Indeed, parking lot trees' leaves were hotter by about 2–4C.

The bottom graph displays a physiological measurement called stomatal conductance — essentially a measure of water flowing out of a leaf's stoma, or the final step in transpiration. Stomatal conductance declined after midmorning for parking lot trees, either in isolation or in a group, but remained higher for trees grouped over turf. Follow-up studies examined actual water loss by trees and more or less confirmed the stomatal conductance results: In some cases there was no significant difference between the asphalt and turf situations, and in others, trees on asphalt transpired less water because they closed their stomata due to the heat.

Parking lot trees have it rough. The long-wavelength radiation coming from the hot asphalt heats the trees up from below and shuts their photosynthetic processes down. Cooling requires mass plantings, perhaps taking away more parking spaces, or perhaps grassy paving (Figure 2.18).

All in all, shade trees can be really shady. One study, performed in Thessaloniki, Greece, indicated that a paper mulberry, *Brussonetia papirifera*, found in an urban park, won the cooling competition: A single tree reduced temperatures by 5–20 degrees Celsius, or 9–35 degrees Fahrenheit![48]

Evapotranspiration is high from watersheds and lawns.

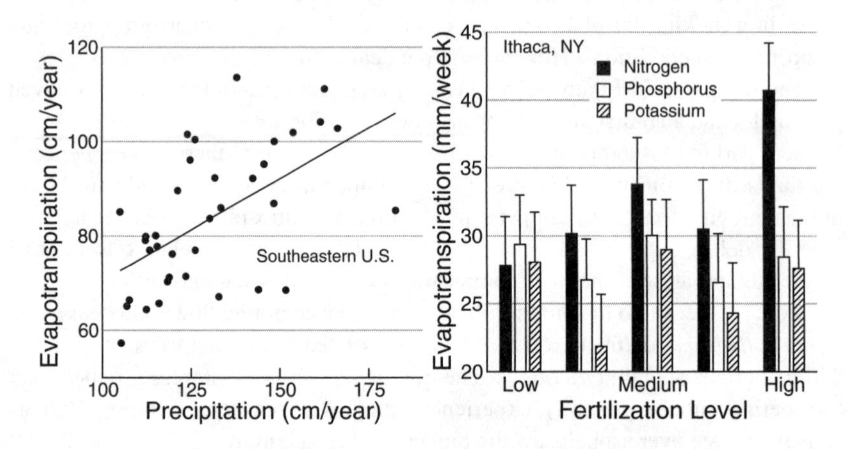

Figure 2.16: The left graph shows that about 60% of the precipitation falling on watersheds in the southeastern United States goes back into the atmosphere through evaporation and transpiration (data from Lu et al. 2003). In essence, these measurements subtract the water flowing out of rivers from the amount falling into those rivers' watersheds as rain. Vegetation transpired away the difference, in combination with simple evaporation. At right, we see evapotranspiration rates from Kentucky Bluegrass "lawns" fertilized at different levels for different nutrients (after Ebdon et al. 1999). Granted, these lawns in Ithaca, New York, were experimental systems, but they evapotranspired about 30 mm of water per week, or 4 mm per day, much more than the amount needed to cool urban asphalt (see Figure 2.10).

As a broad check on the cooling potential for trees and forests, the left graph in Figure 2.16 arises from measures of the precipitation falling into watersheds and the water flowing out of the streams draining the watersheds. Evapotranspiration makes up the lost water.[49] Durham gets, on average, almost 4 inches of rain each month, or about 1/7 inch (3.2 mm) per day. One-third of this amount equals that needed for cooling hot days (compare the numbers in Figure 2.10), if only this rain were somehow stored in the soil instead of diverted to stormwater systems. Here we see that nearly two-thirds of the rain, twice that needed for cooling, is evapotranspired away naturally.

I admit my bias against lawns, in favor of trees, and the results of the right-hand plot shocked me. Those perfect lawns, doused in fertilizers and pesticides, are the antithesis of ecological messiness. But from the perspective of cooling potential through evapotranspiration, a lawn could far exceed a tree. Here I'm not terribly interested in how evapotranspiration changes with fertilization level, but mainly just the average value compared to trees: For comparison, see Figure 2.14, where we calculated that a tree provides about 1 mm of water per day over it's spread. This figure shows 30 mm per week for lawns,[50] about 4 mm per day, working out to 4 kg/day over a square meter.[51] That's much more than the amount needed to cool pavement, and it certainly motivates grassy paving (see Figure 2.18) from a city-cooling perspective.

Realistically, however, environmental features like wind speed obscure these cooling effects, and worse, thermal radiation from buildings can hurt landscape plants. A specific experimental setup in Texas showed that turfgrass evapotranspiration doesn't reduce the outside wall temperature of nearby buildings.[52] In other words, grass lawns don't reliably provide the shade of a tall tree next to a building, and that alone can be important regardless of evapotranspiration levels.

Beyond considering a lawn's cooling benefits, turf perfection comes with a carbon cost from maintenance (Figure 3.10) and water costs for irrigation. For example, water costs outweighed cooling benefits for turfgrass in Tuscon, Arizona, where rocky landscaping with low-water-use shrubs earned the most cost-effective rating.[53] Yet, even while dead, tall grass shades the ground, helping reduce heat island effects by preventing the high thermal mass soil from heating up.

Also remember that trees and grass aren't always exclusive of one another. Trees handle short-term droughts better by using deeper water sources, and many different types of vegetation can be used to lower temperatures for almost any situation.[54] Even painting walls white can be a good thing, lowering temperatures 2C below air temperature, but so could a wall-climbing, wall-destroying ivy.

New developments can plan for shade.

Figure 2.17: At top, outside seating in a small urban canyon known as Brightleaf Square. Trees between the buildings intercept radiant energy between high thermal mass surfaces (compare Figure 2.8 with no interception), as well as provide shade and comfort to restaurant patrons. At bottom, a relatively new gated development arising from warehouse land along Broad Street in Durham, North Carolina, photographed from above in 2005 (courtesy Duane Therriault of City of Durham GIS). Note the spots along the subdivision roads reserved for shade trees.

In the discussion of the urban heat island effect I showed the results of Figure 2.8, indicating that the "urban canyon" explained much of the phenomenon. A photo shown there had two tall buildings with nothing between them but air, and thermal radiation emanating back and forth from one building and absorbed by the other and the street below. The top photo in Figure 2.17 shows how well-placed trees can intercept the radiation bouncing from impervious surface to impervious surface. Slightly lower albedo paving forms portions of the walkway, helping reduce temperatures by reflecting more light. Besides trees, umbrellas also provide convenient shade and have an important role in urban environments. Though the umbrellas are, technically, impervious surfaces, they keep the sun off more thermally massive impervious surfaces that contribute to heat islands. The umbrellas heat up, but much like the leaves on a tree, quickly transfer their heat to the surrounding air. The pavement the umbrellas shade, however, would hold its heat much longer, on into the evening.

Of course, designers of this commercial area had people's comfort foremost in mind rather than optimal traffic flow and maximized building volume. They likely weren't thinking of the public good of reducing the urban heat island: These restaurants need an atmosphere designed for comfort and relaxation, encouraging pleasing social interactions and higher revenues. Later (Figure 5.9) I show that vegetation enhances social activities in public housing, just as it does at these restaurants. Vegetation also correlates with wealth (Figure 6.13), just as it does at these restaurants. Furthermore, it's a fact that cities keep growing along with the human population (see Figure 1.1). Growing a city means transforming forests and farmland into housing developments and amenities, and, by definition, that means the loss of trees. Yet growth need not mean losing the quality of life that comes from living amidst nature, and this private-sector example, if demanded by citizens, could be replicated for the greater social good.

The bottom photo displays a relatively new development from 2001, arising from barren warehouse land previously stripped of trees. Arguably, the high density doesn't help recovery by nature, but at least no trees were harmed in the most recent land-use change. This development has no big trees today, like many others that are mass-graded or built on former agricultural fields or industrial land. Yet, to the developer's credit, it was designed and constructed for future trees that will shade the streets and buildings. Each semicircular unpaved location along the roads and parking spots accommodates a small tree, and, someday, those trees will shade their surrounding impervious surfaces. This development also reflects, perhaps, the underlying reason for the high-income, low-canopy point in Figure 6.13 and provides a perfect example of subdivision age partly predicting the amount of vegetation in a neighborhood.[55]

Paving and grass can be combined.

Figure 2.18: Vehicles need sturdy parking surfaces, implying high thermal mass, and driving on grass lawns compacts the ground and kills the vegetation. Despite the conflict, parking can coexist with lawns. At top are two examples of old, low-tech grassy paving surfaces, one using bricks and the other using holes (photos by Paula Bailey). In both cases, the cement supports the vehicle, and the not-quite-filled holes protect the vegetation from compaction. The bottom photos show an example of modern grassy paving with a plastic substructure providing the vehicular support (photos by Invisible Structures, Inc.). The modern system has very little thermal mass.

Grassy paving (see Figure 2.18) satisfies many needs simultaneously. Its permeability allows water to pass through, reducing stormwater runoff into retention ponds, infiltration trenches, and streams. Filtration at the parking lot source reduces downstream needs for water treatment, particularly when the stormwater becomes drinking water.[56] Reducing stormwater pollution reduces problems with urban streams, including the less abundant and lower diversity of both macroinvertebrates and fish (see Figure 1.10). These problems arise directly from so much stormwater being efficiently and rapidly transported from impermeable surfaces and into stormwater systems, which usually empty directly into urban streams.[57] Grassy paving would, at worst, slow down this transport and, at best, eliminate most runoff and pollutants during light rains.

Impermeable, paved surfaces also reduce on-site water retention, reducing water availability to trees and other vegetation. Grassy paving would increase water infiltration, recharging soil moisture levels, helping trees grow shadier. As we know, grass cools the air through transpiration (Figure 2.16), but the water has to stick around to be transpired. Grassy paving helps solve these problems, while solid, impervious pavement enhances them.

Shade from parking lot trees becomes less important as more and more grass in grassy paving replaces the impermeable surfaces more and more. Certainly, drought conditions would wreak havoc with so much grass: I imagine parking lots of brown by August in Durham, North Carolina, but even brown grassy paving is prettier than hot and dirty asphalt. Other times of the year, however, all those unused parking spaces (see Figure 2.12) might turn green, giving more of the city a pleasant, park-like look that people could enjoy (see Figure 5.9). Grassy paving also demands the right soil. When Durham's soils become wet and saturated, they act like a lump of wet clay. Grassy paving on a lump of wet clay might be too squishy for parking cars.

Let's not forget that grassy paving, being partly alive, needs consistent upkeep, perhaps killing tree seedlings as they recruit into the grass, and some watering during droughts. This maintenance could be costly for a small business, whereas the public gains most of the citywide air and water quality benefits; perhaps some cost-sharing measures could be worked out.

Chapter 3

Energy Use and Carbon Budgets

In this chapter I consider carbon and energy balances within cities and between cities and nature, and the question of the long-term sustainability of these quantities. Americans use a lot of energy, dominated a century ago by coal, then overtaken by petroleum and natural gas, and now, once again, dominated by coal. Even longer ago, trees supplied most of our energy, and recent hopes have pinned the future on biomass fuels. Certainly, modern-carbon energy has advantages over fossil-carbon energy, but our total energy use far exceeds any hopes for a substantial biomass solution.

Energy use varies across the states of the United States, with the highest density states having the lowest per capita energy use for both gasoline and electricity. An energy shortage ought not worry us: We have several centuries worth of coal reserves that could power a new era of coal-fired, plug-in cars.

Photosynthesis makes the link between our energy use, our carbon emissions, and growing vegetation, and, of course, it provided the fossil fuels people use today. I quickly overview how photosynthesis strongly connects atmospheric carbon dioxide (CO_2), water, and light, providing the important context for sustainability calculations. Following on from the examination of sunlight and impervious surfaces producing urban heat islands, I demonstrate how light interacts with trees as part of fixing carbon and shading the ground. Recent emissions from burning fossil fuels over the last 100 years added back carbon sequestered from the biosphere over millions of years. These emissions drove the last century's increase in atmospheric carbon dioxide, and I show both the dramatically strong correlation between CO_2 and temperature over the last 400 millennia, as well as recent warming trends. Several examples show how global warming affects other species directly in the way their lives take place. In addition to global warming, local warming from heat island effects also show direct changes in organism-level properties. These changes take place due to both increased temperatures and increased CO_2 concentrations. Finally, urbanization changes the behavior of soils, or at least the microbial life they contain, increasing respiration over periods of many decades.

A simple carbon footprint calculation for Durham County, North Carolina, shows that even with complete countywide coverage by natural vegetation, citizens' carbon footprints greatly exceed their county boundaries. As for urban vegetation, even though large trees hold large amounts of carbon and grow by many kilograms each year, cities must check that growth to preserve services that citizens demand and limit damages to publicly and privately held assets. Urban trees present maintenance challenges, like a bull in a china shop, demanding careful cost-benefit considerations. This maintenance requires the use of fossil-fuel-powered implements and vehicles, and these carbon costs need accounting in any carbon sequestration calculation.

Leading up to an "energy footprint" calculation, I show that urban trees could play an important role in reducing energy use, but mainly in cooling our houses through shade. Trees also reduce winds, and during the summer, wind reduction increases energy demands. In the case of winter heating, which uses more energy than summer cooling, shade increases energy demands while wind reduction decreases it. These competing trade-offs, balanced differently in different areas of the globe, means that using urban trees for energy reduction needs careful consideration.

U.S. energy sources have changed.

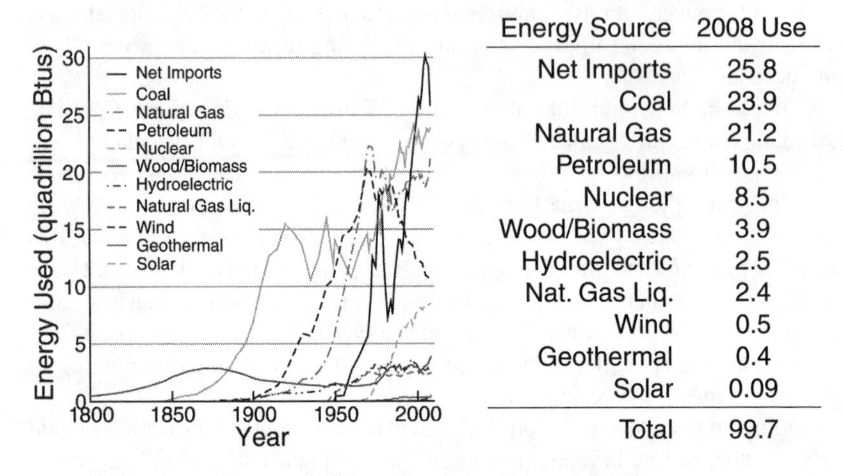

Energy Source	2008 Use
Net Imports	25.8
Coal	23.9
Natural Gas	21.2
Petroleum	10.5
Nuclear	8.5
Wood/Biomass	3.9
Hydroelectric	2.5
Nat. Gas Liq.	2.4
Wind	0.5
Geothermal	0.4
Solar	0.09
Total	99.7

Figure 3.1: Sources of energy used in the United States over the last 200 years. The table shows each source's 2008 contribution to total energy use. Domestically produced petroleum has dropped off lately, but imports have made up the difference (our biggest single supplier being Canada, with Mexico a close second). Domestically, coal and natural gas substitute, in part, for decreasing petroleum supplies. Renewable sources have a long way to go to replace the top four fossil-fuel sources. (Data from Annual Energy Review 2008, U.S. Department of Energy.) Note that a quadrillion equals a thousand million million, or, in scientific notation, 10^{15}.

America's energy demands increased and its energy sources changed over the last two centuries (Figure 3.1). One hundred years ago, all we used was wood and coal.[1] Once industrialization started, coal became our major source.[2] Petroleum and natural gas, with their high energy content (see Figure 3.16), quickly overtook coal, but as those domestic sources become depleted, petroleum imports have made up the loss, and coal once again dominates our domestic energy sources. At present, geothermal, wind, and solar make roughly no contribution.[3] And as we consider retreating back to biofuels, Later I discuss extensively the potential contribution of trees to energy use in Figure 3.16.

Keep in mind that we won't run out of fossil-fuel-based energy anytime soon. The Energy Information Administration (EIA), the official energy statistics supplier of the United States, reports a use of about 1 billion tons of coal per year, with reserves estimated at about 270 billion tons in the United States, and 1,000 billion tons worldwide.[4] Those numbers mean several hundred years' worth of coal to burn, about four times as much as Americans have used over the last century, and lots more fossil-carbon emissions. The EIA also reports a world daily consumption of 83 million barrels of liquid fuels, and total worldwide petroleum reserves as just 1.3 trillion barrels, about 10 years' worth at 100 billion barrels used per year.[5] As dire as that sounds for 20 years hence, many countries invest only enough money in finding reserves to maintain a 10-year horizon.

Averaged over the last century, roughly speaking, all these energy sources add up to 30–50 quadrillion (10^{15}) Btus per year, with present levels at 100 quads per year.[6] Here's an interesting calculation. The United States, including Alaska,[7] covers about 10 million square kilometers, or 10^{13} square meters. Thus, America's annual energy use in 2008 averages about 10,000 Btus/m^2. For comparison, averaged over the entire United States, vegetation has a net primary productivity that works out to maybe 6,000 Btus/m^2 (see the rough calculations discussed in Figure 3.17)![8] Annually, we use nearly twice the energy that all our plants fix in their tissue. Even the 2008 use of 3.9 quads of wood and biomass energy works out to 390 Btus/m^2 each year, more than 5% of the biosphere's annual net primary productivity, but just 3% of our total energy demand. Our use also reduces the availability of plant material to other nonhuman consumers. Most of the atmospheric carbon dioxide fixed by plants feeds into future respiration by other organisms, like bacteria, fungi, and herbivores, that use the energy contained in plant matter and release the CO_2 back into the atmosphere. Using biomass for energy cuts back their consumption. It is a simple fact that biomass will never satisfy our present energy use.

States vary in their gasoline and electricity use.

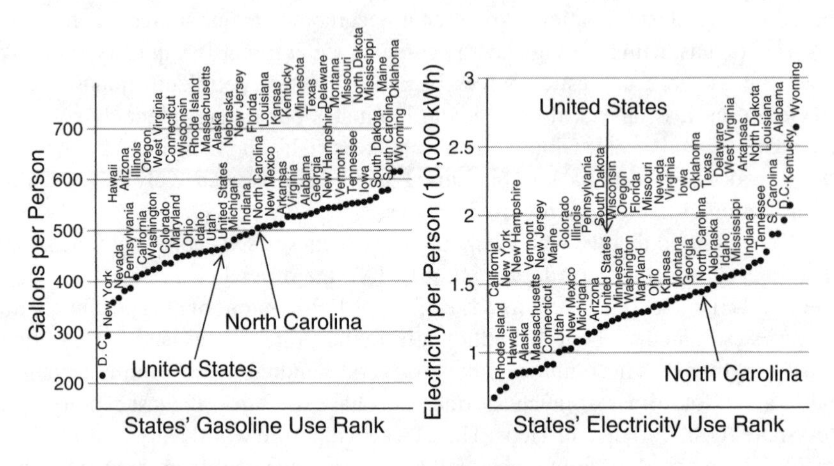

Figure 3.2: Per capita consumption of gasoline, at left, and electricity use, at right, ranked by state-level averages (2003 use data from www.energy.ca.gov). The average American uses about 450 gallons of gasoline, representing about 54 million Btus of energy, and about 12,000 kWh per year, representing about 41 million Btus of energy. Assuming 300 million Americans, these energy uses sum to 30 quadrillion Btus of the total 100 quads for the United States. The remainder comes from commercial and industrial uses.

Per capita gasoline and electricity use (see Figure 3.2) varies widely across states, with California at the low end and Wyoming at the high end.[9] First off, a discussion of gasoline and electricity use requires a short review of energy and power. Applying a force over a distance takes energy, with the Joule, Calorie, and Btu being several such measures.[10] For example, it takes about 10 Joules of energy to lift a kilogram up 1 meter, roughly from your waist to your head.

Power means energy used over a time interval, with the Watt being a common measure. More energy used over the same time interval takes more power, and, likewise, the same energy expended over a shorter time interval. It takes about 10 W of power to apply the force to lift that kilogram up in one second, but it takes less power to apply the same force if you take a longer time to do it. A kiloWatt is 1,000 W, and with that power you could lift 100 kg up 1 m in 1 second, 10 kg up in 0.1 second, or 1,000 kg up in 10 seconds. All three cases require 1 kW of power, but lifting the 10 kg took 0.1 second, requiring 0.1 kW-seconds of energy, the 100 kg took 1 second, requiring 1 kW-seconds of energy, and the 1,000 kg took 10 seconds, or 10 kW-seconds of energy. Your electric company charges you for energy, measured in kilowatt-hours.

One kWh of energy equals 3,400 Btus; the average North Carolinian uses about 14,400 kWh per year, the same as 49 million Btus of electrical energy, and similar to the North Carolinian's 60 million Btus of gasoline energy use. Here's another way to see our energy use. Including Alaska, we have 8.2 acres per person.[11] If I took my 500 gallons of gasoline, at about 2.8 kg (and 120,000 Btus) per gallon of gas, and poured it over "my" 8 acres, there'd be 42 grams/m^2 (roughly three tablespoons), or 1,800 Btus per square meter per year.[12] Again, compare this energy use to the 6,000 Btus/m^2 fixed by plants (Figure 3.17): way too close for biofuels. Likewise, my electricity use works out to about 1,500 Btus/m^2. Commercial, industrial, and other transportation uses make up the remainder reported in Figure 3.1.[13]

How does this energy use compare to sunlight's energy? Every day nearly 28,000 Btus/m^2 of energy strikes the top of the atmosphere, of which 16,200 Btus/m^2 hits the ground,[14] being available for, say, plants or photovoltaic cells. That number comes to 6 million Btus/m^2 each year, compared to our measly annual use of 10,000 Btus/m^2! Given the solar energy striking the ground, about 10 m^2 of 100% efficient solar panels would satisfy a North Carolinian's electrical needs, or more realistically, 50 m^2 of 20% efficient solar panels, an area roughly 15 ft by 30 ft.[15]

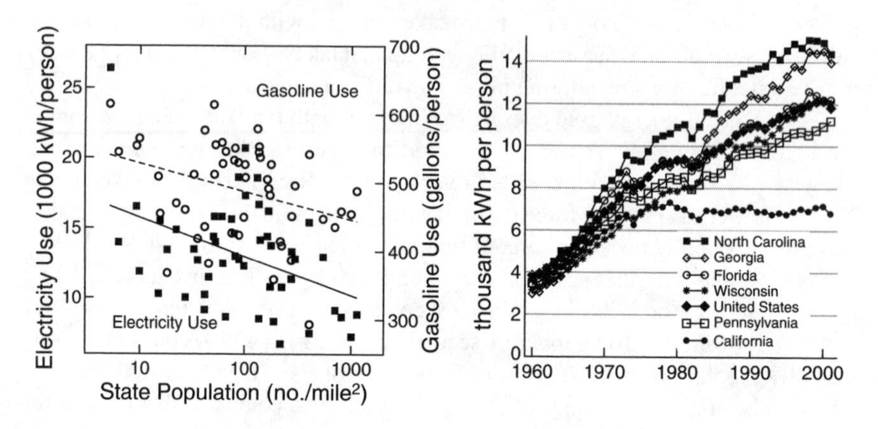

Figure 3.3: At left, gasoline and electricity use as a function of a state's population density (excluding Alaska and the District of Columbia). Cities promote reduced per capita energy use, and, compared to the lowest density states, the highest density states use just 60% of the electricity and 80% of the fuel on a per capita basis. The right plot shows electricity use through time for selected states in the Southeast and California (after North Carolina Sustainable Energy Association).

Urbanization greatly helps us use energy more conservatively. Notice that Figure 3.2 shows how New Yorkers have very low gasoline use, undoubtedly because people in New York County, at 70,000 people per square mile, live without cars. Wyoming has the highest use, undoubtedly because its population is spread out over vast distances. Electricity use also reflects this pattern.

Testing this idea of whether a state's population density influenced energy use, the left-hand graph in Figure 3.3 shows per capita gasoline and electricity use against a state's population density.[16] The fitted lines demonstrate clear trends: People use more energy when they're spread out.[17]

Two obvious mechanisms seem likely. First, people living closer together need less gasoline to go where other people are. Second, the closer people live to one another, the closer people are to electrical generating stations. For example, the low-population density of Wyoming probably means long transmission lines between electrical generation plants and homes, and more energy lost in electrical transmission. Long transmission lines have high line losses, bumping up the per capita energy use of low-density states.[18]

Over the last 50 years, people's electricity use changed, more drastically in some states than others. The right-hand figure shows per capita electricity use in several states.[19] Consider the electricity used in North Carolina — about 14,400 kWh per year. As a personal example, my electric bill shows that over the last year my house has used about 10,000 kWh for *four* people, just one-sixth of the amount indicated per person. However, my electric bill doesn't include the electricity used at my office, my wife's office, my kids' schools, and all of the other places we spend time. That use presumably amounts to 80% of our electrical energy demands. My family meets the household average: The U.S. Energy Information Agency reports that in 2001, 107 million households used 1.1 trillion kWh, working out to 11,000 kWh per household. Electrical energy conservation begins at home, but elsewhere lies the real impact.

By the looks of things, Californians have become energy-misers, but many factors like population density and climate factor in here, too. I'm uneasy comparing energy use between Los Angeles, for example, and agricultural areas of North Carolina. I've lived in both states, and I don't begrudge North Carolinians their air conditioning as their socioeconomic conditions improved, for the health reasons I show later.

Economic productivity correlates with energy use.

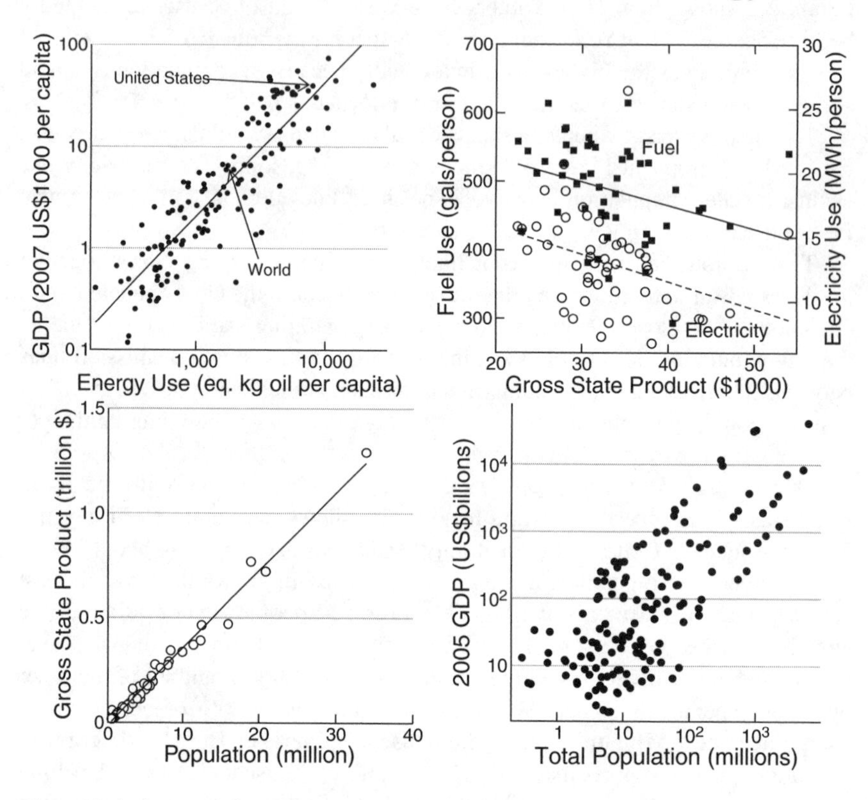

Figure 3.4: At top left, 2004 per capita gross domestic product (GDP) increased with per capita energy use (in oil equivalents) for 144 countries and other entities (data from the World Bank). Within the United States, however, the top right graph shows per capita energy use decreases with per capita gross state product (GSP) (data from the Annual Energy Review 2008). Does energy drive GDP, or does GDP drive energy use, or neither? A near-perfect correlation between GSP and state population size, at bottom left, dictates the striking comparison with Figure 3.3). Globally, population size determines GDP in a much less reliable manner, as seen at bottom right.

We often hear the phrase "energy versus the economy" when people discuss energy-reducing strategies, implying that reducing energy use directly reduces economic activity. Perhaps an argument for reducing our energy use means an argument for reducing our economic activity, thereby reducing our "quality of life." In case the argument for urban trees involves energy reduction, I look briefly at the association between energy and the economy. The top left plot of Figure 3.4 shows per capita energy use and gross domestic product for 144 entities — mostly countries — and the world average in 2004, using data provided by the World Bank.[20] Gross domestic product (GDP) means the total value of goods produced and services provided by property and people located within an entity.[21] According to this World Bank data, the 2004 world average annual per capita energy use is the equivalent of just over 1,790 kg of oil per year, and the average per capita GDP is $US6,500. Assuming 120 kg per barrel of oil and multiplying this energy use by 6.6 billion people gives about 100 billion barrels, or about 270 million barrels per day. Iceland has the highest per capita energy use, the rightmost point, and Luxemborg has the highest per capita GDP, the uppermost point.[22] This plot supports the broader point that more robust economies use more energy. Or is it that using more energy yields a more robust economy? If the latter, scaling back energy use seems to imply scaling back the economy.

Do I believe the energy numbers? These data state the U.S. per capita energy use as roughly 8,000 kg per year, measured in oil equivalents. Assuming 120 kg per barrel of oil, and that a 42 gallon barrel of oil yields about 10 gallons of gasoline, means that 8,000 kg of oil gives about 2,000 kg of gasoline. At 2.8 kg per gallon, that equals 700 gallons of gas (or about 2 gallons per day). This number agrees roughly with the 450 gallons per person we've already seen in Figure 3.2, so, yes, I believe the numbers.

Does the same correlation hold within the United States? Nope. At top right I plot the 2005 per capita gross state product[23] (GSP) against the state-averaged per capita energy use numbers from Figure 3.2. Gross state product, like GDP, marks the value of goods and services produced within a U.S. state. Across these states, the saying should be "energy or economy," even though across countries the saying seems to be "energy and economy." However, the bottom left plot shows that this between-state comparison simply reflects the very strong correlation between a state's GSP and its population, and the bottom right plot demonstrates the similar connection between the country-level quantities.[24]

Future economic productivity increasingly confronts the concerns of energy sustainability and global environmental issues, which come together through the solar power that drives plant growth. We cover that topic next.

Photosynthesis links carbon, water, nitrogen, and sunlight.

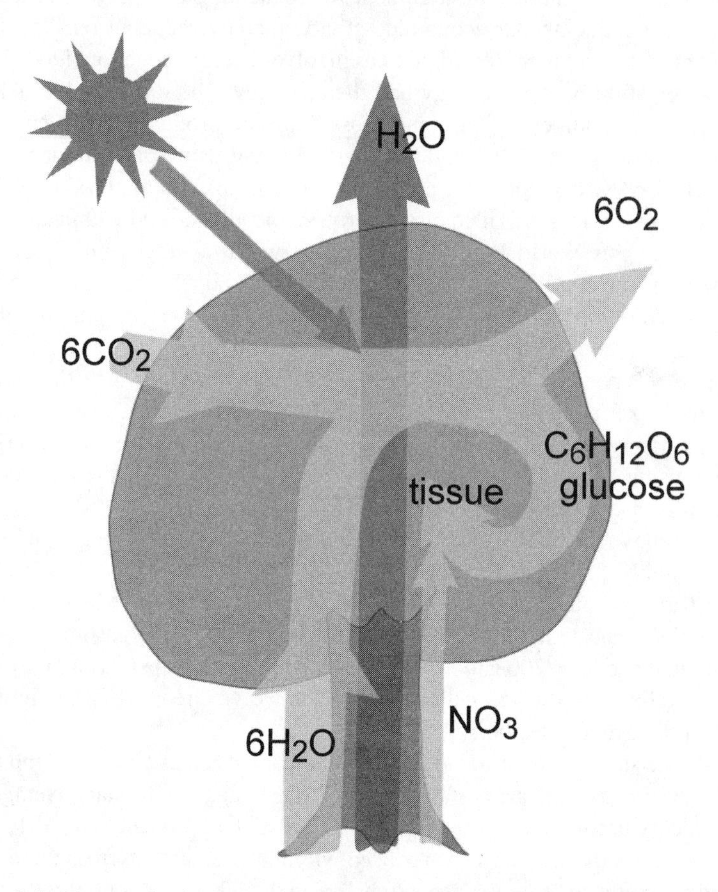

Figure 3.5: Plants take in carbon dioxide (CO_2) from the atmosphere and water (H_2O) from the soil, and use solar energy to produce a variety of energy storage molecules, including the sugar glucose ($C_6H_{12}O_6$). Behind much of these photosynthetic processes, plants transpire lots of water to transport soil nutrients like nitrogen (NO_3^-) to build proteins.

Our lives completely depend on photosynthesis. Figure 3.5 shows its essential features, linking many features I discuss later. Photosynthesis pulls carbon dioxide from the atmosphere and water from the ground, transforming sunlight into sugar to store energy. Plants metabolize the sugar — just like their herbivorous and frugivorous consumers — and recover the energy, using it to make all sorts of biological material, including the volatile organic compounds (VOCs) emitted along with oxygen, O_2. Besides solar energy and carbon dioxide (CO_2), tissue-making depends on many other chemical compounds and elements. Pulled up along with transpired water from the soil, most plants need, for example, nitrate, NO_3^-, to build nitrogen-containing amino acids, the building blocks of proteins, and nucleotides, the building blocks of RNA and DNA.[25] By dry weight, plants comprise about 44% carbon, 44% oxygen, 6% hydrogen, 1–4% nitrogen, 0.5–6% potassium, 0.2–3.5% calcium, 0.1–0.8% phosphorous, and then a slew of minor but often essential elements.[26] In the end, we eat and burn these sugars and tissues, making our lives completely dependent on past and present solar power, save for small nuclear energy contributions.

Plants recycle lots of carbon in a very short amount of time.[27] The atmosphere contains around 750×10^{15} grams — 750 trillion kg — of carbon, but in a single year plants fix roughly 120 trillion kg into their tissues and oceans absorb 90 trillion kg. Plants, animals, fungi and bacteria consume living and dead tissues of plants, and, for the most part, respire an equal amount of carbon back into the atmosphere, offsetting the carbon fixed by vegetation. That's a huge amount of recycling: A carbon atom, on average spends only about four years in the atmosphere (750 trillion kg)/(210 trillion kg/year) before, with a good chance, being pulled back into some plant's stomata and fixed into its living tissues. On top of these amounts, humans put 6×10^{15} grams of fossil-fuel carbon into the atmosphere each year, a mere 6 trillion kg, or just 5% of the natural fixation by plants.

The biosphere buries just a small fraction of its vegetation. Multiplied over hundreds of millions of years, this buried plant material amounts to significant deposits of hydrocarbons. Over the last century or two (see Figure 3.1), we used fuels from these buried and concentrated fossil plants, releasing carbon long ago pulled from the atmosphere and sequestered out of the biosphere. It's ironic that we're trying to both use the solar energy stored in those fossil hydrocarbons and subsequently sequester the released carbon dioxide. If we just stopped using the solar power contained in these fossil fuels in favor of direct solar power, we could stop worrying about carbon sequestration.

Atmospheric CO_2 increased with human emissions.

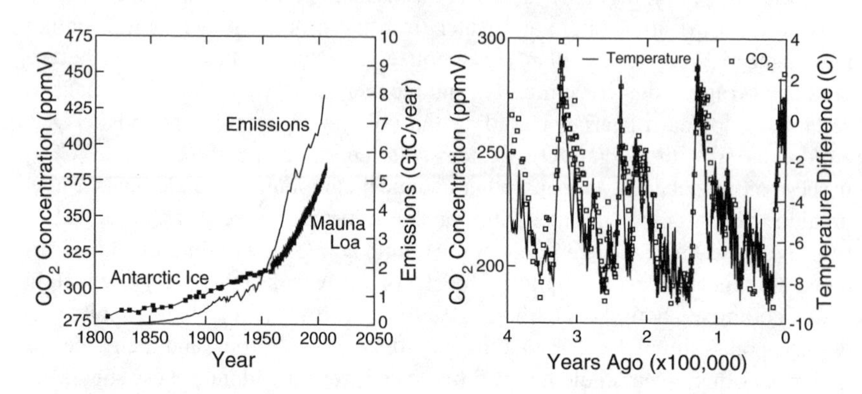

Figure 3.6: The left plot extends the recent CO_2 concentration plot to include the last two centuries and shows emissions from fossil fuel, cement, and gas flare sources combined (data from Marland et al. 2008). The steep change in the two curves' slopes as they shoot upward take place remarkably close in time, and total emissions roughly equal the total increase in atmospheric carbon. Shown at right are CO_2 levels (the lines) from the last 400,000 years, along with corresponding temperatures (the squares), measured from air bubbles in 3.6 km deep glacial ice cores from Vostok station in the eastern Antarctic (after Petit et al. 1999). Variations in CO_2 levels and temperature clearly reflect one another, and our recent fossil fuel CO_2 emissions likely signal future (and present) temperature increases.

Simple facts include that the world is round, the Earth orbits the Sun, natural selection drives evolution, and CO_2 levels are incredibly high. Despite sunlight-driven photosynthesis absorbing lots of CO_2, atmospheric CO_2 levels clearly and undeniably increased over the last 200 years. I show the recent trend in the left graph of Figure 3.6. When I was born in 1960, the level was about 320 ppmV (parts per million by volume), and about 50 years later the levels are more than 380 ppmV, nearly 20% higher. For convenience, I've left space on the right-hand side of the graph to fill in the curve over the next four decades. Scientists don't like projecting curves very far beyond their data: What if an asteroid hits? Or a mega-volcano erupts? Or the oceanic currents change? Or the frozen tundra emits more carbon than estimated when it thaws? Many things could cause CO_2 predictions to be off, but, still, just guessing here, CO_2 levels are likely to be somewhere around 460 ppmV when my children reach my age today.

People cause rising CO_2 concentrations. A rather striking correlation, also shown in the left-hand plot, exists between the very recent, very rapid rise in CO_2 and the simultaneous, very recent, very rapid increase in CO_2 emissions by people.[28]

It's also a simple fact that temperature and CO_2 levels show a high correlation over a very long time, as the graph on the right shows. The lines depict CO_2 levels, and the data points indicate temperature — it's not simply a matter of lines connecting points; they're two different things! Physical processes tightly correlate the two, a correlation much stronger than, say, asthma and air pollutants (see Figure 6.4).

Again, these graphs show that present CO_2 levels are passing through 380 parts per million, compared to the highest historical level over the last 400,000 years, which saw a maximum of about 300 ppm.[29] Since then, we (me included) have sent a great deal of carbon dioxide into the atmosphere. A carbon atom spends about four to five years in the atmosphere before some plant fixes it into its tissue or it gets mixed into ocean water.[30] If we could just stop emitting all fossil carbon, estimates show that the biosphere could pull atmospheric CO_2 levels back toward "equilibrium" in about 50 to 200 years.[31]

Are humans responsible for increasing CO_2, or is it just "natural" variation having nothing to do with humans?[32] One study examined this long time series of data, identifying the time scales present in CO_2 changes. Important scales are daily (night versus day levels), annual (summer versus winter levels),[33] and glacial scales of around 100,000 years.[34] Besides these, the last 100 years emerge from background CO_2 variation as a unique and strong peak, which gives strength to the argument that the last 100 years are due to humans.[35]

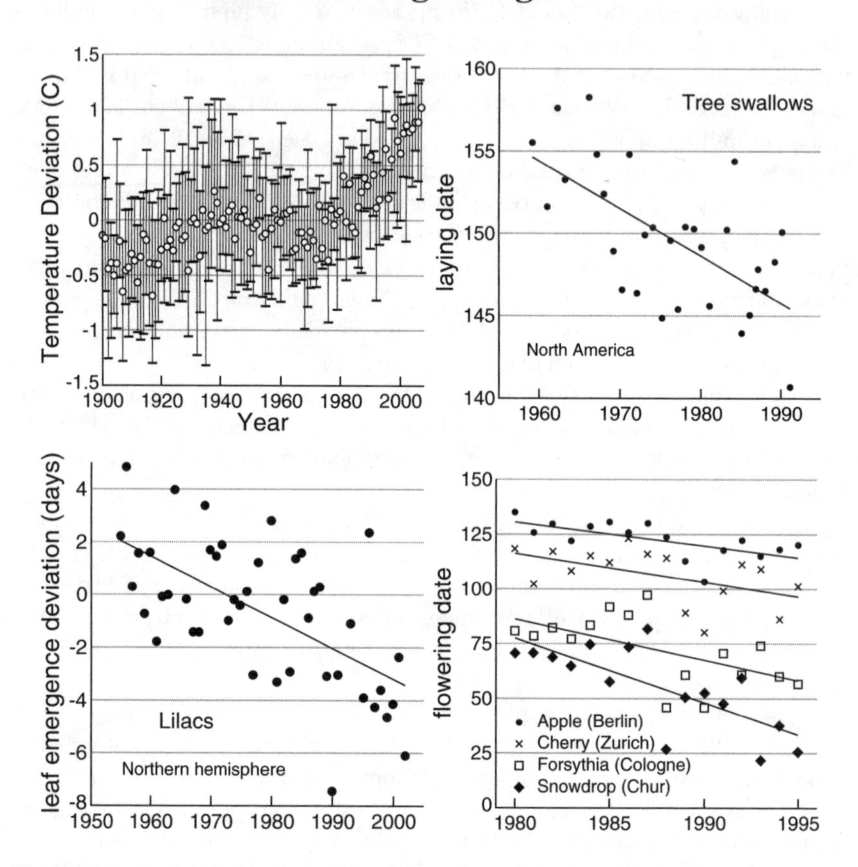

Figure 3.7: At top left, the graph shows the relative temperature change over the last century (after Brohan et al. 2006). The top plot measures a progressively earlier laying date for tree swallows across North America, amounting to about 10 days' change over three decades (after Dunn and Winkler 1999). At bottom left, lilac leaves now emerge in the northern hemisphere about five days earlier than they did five decades ago (after Schwartz et al. 2006). At bottom right, several plant species in Europe flower one to three weeks earlier in the spring than they did just 15 years earlier (bottom right; after Roetzer et al. 2000).

We've just seen that temperature strongly correlates with CO_2 levels, and increasing CO_2 levels correlate with emissions (Figure 3.6). In the top left graph of Figure 3.7, I show the recent increasing global temperatures, which represents just one aspect of climate change,[36] to consider alongside the other three plots showing ecological changes. These plots examine the timing of life events, called phenology, caused by increased temperature and carbon dioxide.[37]

At top right, a massive study of almost 3,500 tree swallow nests, using data reported by a large network of ornithologists in North America, showed that the laying date has changed by nine days over a time span of a few decades. In this case, scientists suspect that warmer temperatures cause insects to emerge earlier in the season, the tree swallows eat these earlier available insects, obtaining enough resources in a shorter time, allowing them to lay their eggs earlier in the season.[38]

This earlier laying date might not be so bad for adult tree swallows, but this same effect for *Parus major*, a common British bird called the great tit, creates a problem because the earlier laying date also means an earlier hatching date. At that point it's up to the adults to feed their offspring, but the chicks' food source doesn't change its development time. That leads to a mismatch between when the young hatch and when their food supply appears. That's a problem for offspring survival.

Climate change also affects trees and shrubs. Measurements taken from three lilac species, *Syringa chinensis*, *S. vulgaris*, and *S. oblata*, across the northern hemisphere clearly demonstrate, at bottom left, that the date that the first leaf appears on the shrub moves up by a little more than one day each decade. Where these shrubs grow, the last spring freeze date moves up even faster, about a day and a half per decade.[39] Yet another climate change study, at bottom right, shows that the flowering dates of several tree species in Europe, along with a flowering bulb (the snowdrop), echo the lilac results.[40]

For the climate change skeptic, perhaps all these ecological changes take place because scientists study systems close to growing cities having increasing heat island effects (see Figures 2.7 and 2.7). Perhaps these changes aren't due to global warming, or rather it's all due to urban heating. Scientists thought about that idea, as I show next.

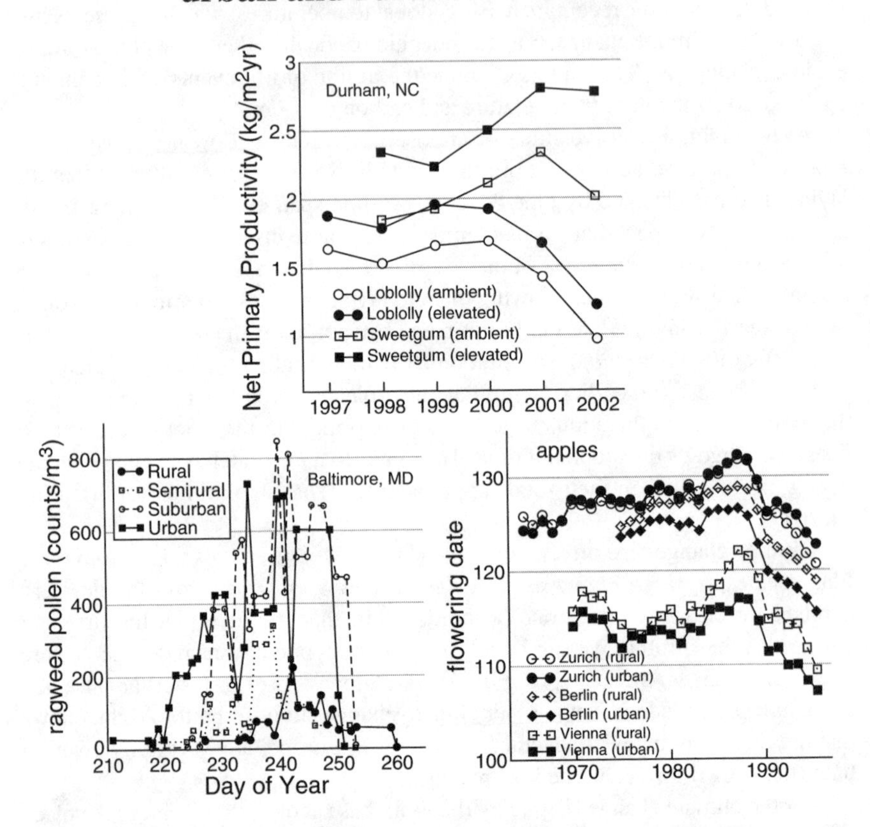

Figure 3.8: At top, net primary productivity increases, at least in the short term, for Durham, North Carolina, loblolly and sweetgum forests grown with carbon dioxide elevated above ambient concentrations (after DeLucia et al. 2005). In the bottom left plot, data taken near Baltimore, Maryland, show that the production of ragweed pollen, a common allergen, increases with greater urbanization (after Ziska et al. 2003). The bottom right graph shows flowering times of apple trees across central Europe; for the most part, urban areas have earlier flowering dates than so-called rural areas (after Roetzer et al. 2000). Temperature differences between rural and urban environments might mimic the changes due to global warming.

High levels of CO_2 affect plant physiology. Enhanced CO_2 levels increase net primary production (NPP) in both loblolly- and sweetgum-dominated forests (two southeastern trees), as the top plot of Figure 3.8 shows.[41] In these studies elevated treatments increased CO_2 by 200 parts per million by volume (ppmV). Dry matter sequestered by these forests increased by 14–26% in loblolly pine forests and 16–38% in sweetgum forests. These sequestration numbers refer to increased biomass, of which about half is carbon. For example, compare the maximum plant growth values of 1 kgC/m^2/year in Figure 1.2. Young trees, like these measured here in early successional forests, also grow very fast, which means higher productivity.

Unfortunately, tree growth won't solve our global warming problems. Earlier I showed how NPP also increased with water (Figure 1.2) and nitrogen (Figure 1.6). All of these results correlate with one another; for example, greater NPP through greater nitrogen availability demands more water.[42] We ought not expect trees to grow twice as fast and twice as big just because we doubled carbon in the atmosphere. Trees need other things to grow, too. People can expect at best only short-term NPP increases with increased CO_2 because other features, like nitrogen supply, become the limiting resources for plant growth.[43]

Indeed, the urban heat island poses an interesting scientific opportunity: predicting what might happen with global climate change by observing what happens with local climate change. Right between urban and rural areas, with remnant patches of local ecosystems, the urban heat island might mimic the heating of human-induced climate change. What happens there now might reflect what happens globally in our future, warmer world. At bottom left, the plot compares ragweed pollen production in urban Baltimore, Maryland, with increasingly rural settings. Urban and suburban environments yielded much more ragweed pollen. Local warming also reflects dire global warming scenarios: Urban Baltimore CO_2 levels are 30% greater, and its temperatures are about 3.5F higher.[44] Apparently, we can look forward to more ragweed pollen in a warmer urbanized world.

Figure 3.7 demonstrated advancing flowering times in Europe, presumably due to global warming. The bottom right graph shows advancing flowering times for apple trees, one of four species showing similar results, in rural and urban settings. Most of the urban stations show flowering earlier than their paired rural station, except for the Zurich case in which the rural station sits close to an urban area.[45] In addition to earlier flowering, results indicate there may be other, more negative, consequences, for example, mismatches in flowering and pollinator availability, and outright flowering failure.[46] Clearly, changes in both global and local climates affect species other than people, in both urban and rural areas.[47]

Soils contribute to carbon budgets.

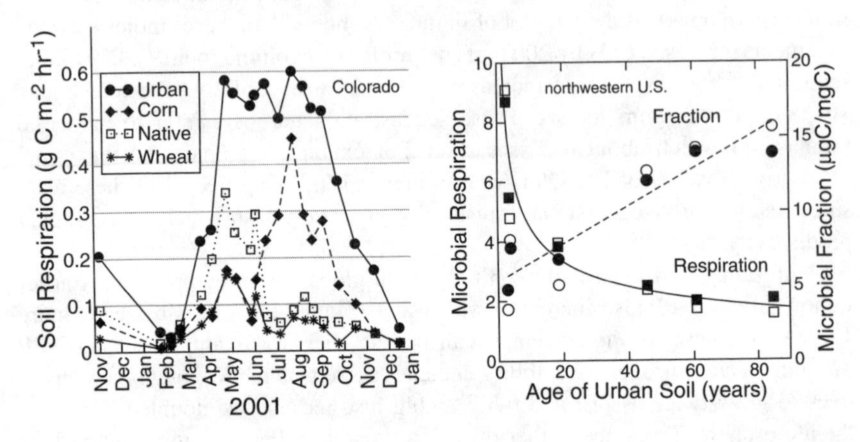

Figure 3.9: The left plot compares soil respiration (CO_2 released from microbes in a soil sample) from native grasslands, wheat fields, and irrigated and fertilized corn fields and urban lawns, using three samples from each type from northern Colorado. Urban soils release large amounts of CO_2 through enhanced respiration (after Kaye et al. 2005). At right, urban soils take time to recover, from the microbial perspective, after "urbanization" (after Scharenbroch et al. 2005). These samples came from Moscow, Idaho, and Pullman, Washington, in a variety of urban situations in years 2002 (filled symbols) and 2003 (open symbols). Recently urbanized soils have relatively high respiration in response. Older urban soils are better developed, having greater microbial carbon content with longer times since urbanization. Microbial activity roughly represents μg respiration of CO_2 per hour per mg of microbial carbon in a soil sample.

Lest we forget what all these plants rest on, climate change involves soils. Ecological communities within soils are part of the carbon cycle, returning carbon to the atmosphere by breaking down dead plants through respiration. Typically, one measures how much CO_2 comes out of a soil sample, reflecting the amount of microbial activity taking place in the sample. In the end, urbanization affects soil ecology.

In the left plot of Figure 3.9, soils from wheat and corn fields are compared with native grasslands and urban environments.[48] Within the urban treatment itself, three urban sites were examined: one institutional, one rental dwelling, and one owner-occupied house. Kentucky bluegrass, *Poa pratensis,* dominated lawns at all three sites. Urban sites were also irrigated, fertilized, and surrounded by trees. In addition to higher respiration rates, urban soil samples have much greater organic carbon and total nitrogen. Furthermore, urban environments had much greater net primary production than the other sites (except for corn), and a much greater allocation of carbon to below-ground biomass.

As an order-of-magnitude calculation, let's say respiration by urban soils amounts to about 0.25 grams of carbon per square meter per hour averaged over the day and year. For the entire year, this value gives a total respiration of about 2 kg of carbon per square meter per year. If we were expecting a balance between plant growth and soil respiration, this number exceeds NPP values in Figure 1.2. Even compared to fossil-fuel emissions, land-use change contributes a lot of carbon to the atmosphere. Through burning fossil fuels, humans have released 270 ± 30 petagrams (Pg; 10^{12} kg) of carbon. At the same time, transformation from rural to urban land use increased soil respiration to the tune of 136 ± 55 Pg. Of this urban respiration, 78 ± 12 Pg have been released from the dark, organic matter stored in the ground, degrading urban soils.

However, urban soils recover from the developmental insult of removed vegetation and topsoil, compression from heavy vehicles, and plantings of grass. Shown in the right plot, results from two northwestern states, Idaho and Washington, demonstrate that, in response to urbanization, microbial activity was initially quite high, gradually reducing to levels seen in other environments.[49] At this later stage, soil respiration balances net primary production inputs through dead organic matter. Besides a calmed-down microbial activity as urban soils age, organic carbon increases, and the results here show that microbial fraction increases, too.

All in all, humans can't forget soils when dealing with increasing temperatures and CO_2 concentrations: It's estimated that with improved soil management soils could sequester 0.6 to 1.2 Pg C per year.[50]

Vegetation stores and sequesters carbon.

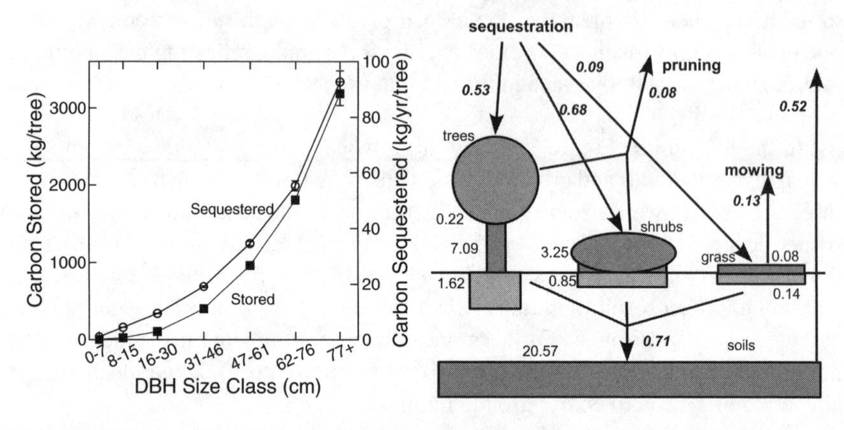

Figure 3.10: At left, the graph shows carbon storage and sequestration for Chicago, Illinois, urban trees, mostly elms, maples, ash, and poplar, of various sizes (data from Nowak 1994). For example, a tree between 47 and 61 cm diameter-at-breast-height holds nearly 1,000 kg of carbon, and increases its carbon mass by about 35 kg each year. Estimates here assume carbon mass is half the total biomass. At right, the diagram depicts the pools and annual fluxes of carbon in vegetation, as well as carbon costs for maintenance (after Jo and McPherson 1995). The pools are in units of kgC/m^2, and the fluxes are kgC/m^2/yr. According to these estimates, trees and shrubs have a positive net carbon sequestration because $0.53 + 0.68 > 0.08$, whereas mowing grass results in a negative net sequestration because $0.09 < 0.13$.

Let me clarify the difference between "stored" carbon and "sequestered" carbon in vegetation. Stored carbon represents present-day mass: Chop down a tree and you get a certain amount of firewood. Sequestered carbon represents how much a tree will grow from one year to the next: If you chop down the tree next year, you'll get a bit more firewood than you would this year.

From a detailed study of urban trees in Chicago, Illinois, for example, trees with 47–61cm diameter (my elbow-to-fingertips measures 48 cm) sequestered 35 kgC/year, and trees with 62–76 cm diameter (my shoulder-to-fingertips measures 79cm) sequestered 55 kgC/year.[51]

Not all urban vegetation has equal carbon sequestration costs and benefits. Results depicted in the diagram in Figure 3.10 summarize measurements for grass, shrubs, and trees, from two residential blocks in Chicago.[52] Connecting the graph and diagram, the crown diameter from Figure 2.13 for open-grown trees of sizes 47–61 cm diameter and 62–76 cm diameter are about 12 m and 14 m, or about 113 m^2 and 154 m^2, respectively. Sequestration values then run 0.31 kgC/m^2/year and 0.36 kgC/m^2/year, close to the 0.53 kgC/m^2/year quoted here. Similarly, carbon storage values run 8.5 kgC/m^2 and 11.7 kgC/m^2, respectively, close to the total 8.9 kgC/m^2 itemized for trees. In Figure 3.12 I'll examine carbon costs associated with maintaining trees.

I've also discussed lawns. Grass needs mowing, for example, only to keep trees from colonizing in the southeastern United States. By itself, these numbers show that mowing with a gas-powered mower puts more fossil-fuel carbon into the air than the grass sequesters out of it. Trees and shrubs, on the other hand, capture more carbon than their required maintenance releases.

Besides urban Chicago, carbon sequestration rates of golf courses in Colorado and Wyoming annually sequester about 1 ton per hectare (about 0.1kg/m^2) during the first 25 or so years after construction, but then reaches a balance when respiration by soils equals sequestration by the grass. Importantly, these results did not account for fossil-fuel use in golf course maintenance.[53] For comparison, the urban Chicago results estimated mowing at 0.13 kg/m^2, more than the golf course grass sequestration above. If those numbers applied to the golf courses, their fairways cost carbon.

In any event, always keep in mind my assertion that all carbon sequestration numbers are absurd! The concept, if it has any validity at all, applies only in the very short term: Herbivory and decomposition release nearly all of the sequestered carbon back into the atmosphere. Only active removal from the biosphere, such as burial in a dump, truly sequesters carbon over the long term. This removal doesn't include burning firewood and laying mulch onto your yard.

Urban pruning can be very intensive.

Figure 3.11: A tree that has undergone extensive pruning, sacrificing shade and beauty, for the sake of protecting power lines during hurricanes. Another photo shows growing damage to a curb and stone wall, not to mention challenges to pedestrians. What good are these urban trees?

We all understand the differences between a tree growing peacefully and unabtrusively in a forest and a tree growing in the city near sidewalks and foundations, on top of sewer lines, and underneath power lines. Here in Figure 3.11 I picture two examples of problematic urban trees. During the summer of 2007, some serious tree pruning took place on Durham's Broad Street before nearby residents complained. City leaders reexamined the work, finding that it wasn't taking place according to stated principles of urban forestry.

No doubt about it though, serious interference issues exist between the trees and the overhead power lines in the top photo. In its unpruned state, if a hurricane happened to come along the tree would take all the power and phone lines down with high repair costs. Even just a falling, rotted branch would cost serious money, not just to clean up the branches and repair a vehicle, but also repair the power line and increase electricity costs. Not only do these pruning costs get passed down to taxpayers, but if the power goes out consumers bear the cost of replacing all the spoiled food in their refrigerators. The bottom photo shows the potential destruction of roots located near roads and walls; we've all seen examples of sidewalks buckled by growing trees. In the wrong places, urban trees require pruning and maintenance, and these pictures show the result.

Human costs also factor in. The National Institute for Occupational Health and Safety reports that between 1980 and 1989, at least 207 deaths were associated with tree trimming, including 68 deaths by electrocutions (contact with power lines) and 52 deaths by falls.[54] Planting more trees beneath power lines means more required trimming, and, inevitably, more fatalities. This 1980s fatality rate means, roughly, that a city of a million people should expect one tree-trimming fatality per decade. Accidents happen, and urban citizens must recognize the complicated balance between city services and trees beyond just carbon and energy. The right tree in the right place needs less maintenance, providing benefits at lower cost.

With these pictures in mind, I roughly divide the remainder of this chapter into two connected parts concerning urban trees, the first concentrating on a carbon budget and the second focused on an energy budget. My carbon budget examination makes little mention of energy and energy sources; rather, it mostly looks at the carbon costs of maintaining urban trees. The second part, urban trees' energy budget, looks at the potential energy savings from well-placed trees, as well as the potential for trees to serve as an energy source.

Carbon costs of landscaping machines are high.

Pruning times (hours)

Tree size (DBH cm)	2.3 hp Chainsaw	Bucket truck	Chipper
1–6	0.05	NA	0.05
7–12	0.1	0.2	0.1
13–18	0.2	0.5	0.2
19–24	0.5	1.0	0.3
25–30	1.0	2.0	0.35
31–36	1.5	3.0	0.4
36+	1.5	4.0	0.4

Removal times (hours)

Tree size (DBH cm)	2.3 hp Chainsaw	3.7 hp Chainsaw	Bucket truck	Chipper	Stump grinder
1–6	0.3	NA	0.2	0.1	0.25
7–12	0.3	0.2	0.4	0.25	0.33
13–18	0.5	0.5	0.75	0.4	0.5
19–24	1.5	1.0	2.2	0.75	0.7
25–30	1.8	1.5	3.0	1.0	1.0
31–36	2.2	1.8	5.5	2.0	1.5
36+	2.2	2.3	7.5	2.5	2.0

Equipment Carbon Costs

Implement	Average C emission (g/hp/hr)	Total C emission (kg/hr)
Bucket truck	147.2	3.2
Backhoe	147.3	5.3
Chainsaw (<4 hp)	1,264.4	1.5
Chipper/grinder	146.4	5.4

Figure 3.12: Maintenance costs associated with trees (after Nowak et al. 2002b). For example, a 31–36 cm DBH tree costs about 17.2 kg of carbon emissions for pruning, calculated by adding use of chainsaws (1.5hr)(1.5kgC/hr), bucket truck (4.0 hr)(3.2 kgC/hr), and chipper (0.4 hr)(5.4 kgC/hr). This 17.2 kg cost nearly equals the annual carbon sequestered by 31–46 cm DBH trees, 19.1 kg, shown in Figure 3.10.

Burning fossil carbon releases carbon sequestered for millions of years. Recapturing this released carbon — sequestering carbon from the biosphere's carbon cycle — has gained recent interest. If city dwellers hope that city trees sequester city-spewed carbon, then city tree carbon costs also need accounting. City trees need pruning and removal, and these actions require chainsaws and other fossil-fuel-powered tools, causing carbon emissions over the long term. How much can power-tool-dependent urban trees help in the short term? Using the tables presented in Figure 3.12, one can consider carbon sequestration that trees provide versus all of these maintenance costs required in urban areas to determine the net carbon sequestration achieved by planting trees of various species. Shown are the carbon costs of city-maintained trees, detailing required times for various powered equipment to prune and remove trees of different sizes.

As an example, let's consider a tree of size 31 to 36 cm DBH. According to these numbers, pruning costs about 17.2 kg of carbon, and complete removal costs about 43 kg of carbon. Compare these numbers with annual sequestration rates of 19.1 kg of carbon and total carbon storage of 400 kg for 31 to 46 cm DBH trees in Figure 3.10. Keep in mind the shading and cooling benefits of urban trees, but if an urban tree needs significant pruning and, in the end, careful removal, its carbon sequestration role doesn't look great. Of course, cities don't prune every tree every year, so these numbers overstate an urban tree's annual carbon costs.

Carefully taking into account these costs and benefits, the study from which these numbers came calculated tree-species-specific breakeven times — a measure of when carbon used to maintain a tree exceeds the carbon sequestered by the tree.[55] Breakeven times vary between species because their growth habits differ, and the rank ordering of tree species reveals good and bad tree species for pulling carbon out of the air. The winners? Tulip trees (*Liriodendron tulipifera*) and white oaks (*Quercus alba*)! I love tulip trees because they grow so fast, providing shade very quickly with low biogenic emissions (see Figure 4.3). At the opposite end, river birch (*Betula nigra*) and pin cherry (*Prunus pennsylvanica*) perform poorly.

I reemphasize that bacterial and fungal decomposition releases *all* of the carbon sequestered by trees back into the atmosphere, even without considering the maintenance carbon costs tabled here. That means carbon sequestration is somewhat mythical. Pruning and tree removal happen because humans want to control the fall of branches and the trunk upon tree death, hoping to avoid greater economic consequences. Only by burying dead trees deep underground, where fossil fuels come from, can we truly sequester a tree's carbon for a significant time span.

Durham citizens export their carbon sequestration.

Carbon Budget

Per Capita Emissions

Carbon from gasoline: 2.4 kg/gallon
460 gallons/year = 1,100 kgC/year
Carbon from electricity: 0.1 kg/kWh
12,000 kWh/year = 1,200 kgC/year

Sequestration

Carbon absorbed by trees
0.43 kg/m^2/year for U.S.
vegetation absorbs about 1,740 kg/acre

Carbon offset

Fossil-fuel emissions offset by tree fixation
(2,300 kg/person/year)/(1,740 kg/acre)
= 1.32 acres/person/year

Area of Durham County: 186,000 acres
Population of Durham County: 225,000 people
(186,000 acres/225,000 people)
= 0.83 acre/person

Figure 3.13: A calculation of Durham County citizens' carbon footprint. On a per capita basis, carbon emissions represent 1.32 acres of net primary production by plants, but the county only has 0.83 acre on a per citizen measure. These numbers underestimate total U.S. per capita carbon emissions; using numbers from Figure 3.1 yields about 10,000 kg per person, not 2,300 kg.

Carbon footprint calculations aren't rocket science. The more you fill your gas tank, the more you fly, the bigger your house, the colder your air conditioning setting, the higher your heating setting, the more goods you buy, the bigger your carbon footprint.

Here's an approximate carbon calculation for people living in Durham County, assuming average gasoline and electricity use (see Figure 3.2). In my calculation, the question I'll ask is how many acres of vegetation do I need to pull the carbon I've used, in terms of my fossil-fuel use, back out of the atmosphere? I'll call that a carbon offset calculation, telling us how our fossil-carbon use compares to carbon use by the biosphere.[56]

Calculating carbon emissions from gasoline requires simple multiplication of gallons used and carbon content per gallon, but emissions for electricity in the Durham area depends on the mix of energy sources, either coal, natural gas, or nuclear. Typically, carbon emissions for electricity generation range from about 0.04–0.13 kgC/kWh, and I'll assume 0.1 kgC/kWh.[57] As the table in Figure 3.13 shows, the average Durham resident each year emits 2,300 kg of carbon, about 2.5 tons.

Just to make the calculation simpler, suppose that every acre of Durham County has trees fixing carbon.[58] Using growth values of about 20 kg/tree/yr for 30–45 cm DBH trees (see Figure 3.10) and crown width and area (see Figure 2.13) of 10 m and 80 m^2, respectively, gives about 0.25 kgC/m^2. This number agrees well with the net primary productivity values of Figure 1.2, with a United States average of 0.43 kgC/m^2, which I use here. Other estimates for eastern and northeastern forests range upwards of 1 kgC/m^2 per year.[59] At the low end, the arid shrublands of the southwestern United States fix less than 0.1 kgC/m^2 per year.[60] Using the United States average, each acre of forested land absorbs about 1,740 kg of carbon, meaning that the average Durham resident needs about 1.3 acres of forest to offset their emissions, an area nearly twice the per capita area of the county.[61]

Whatever the detailed assumptions,[62] it is clear that, on average, citizens of Durham County emit more carbon from their fossil fuel use than the trees in their county can absorb. In this sense, Durham residents export their carbon sequestration needs somewhere else, relying on some other place's trees to balance atmospheric carbon levels.[63] Of course, most Americans are just like Durham's residents, and that's why atmospheric carbon dioxide levels are rising. We're using fossil carbon beyond the ability of the biosphere to counteract it, and whose job is it to make up the difference? Perhaps areas that have positive sequestration could be compensated for that service to urban residents?

Trees and white paint reduce energy consumption.

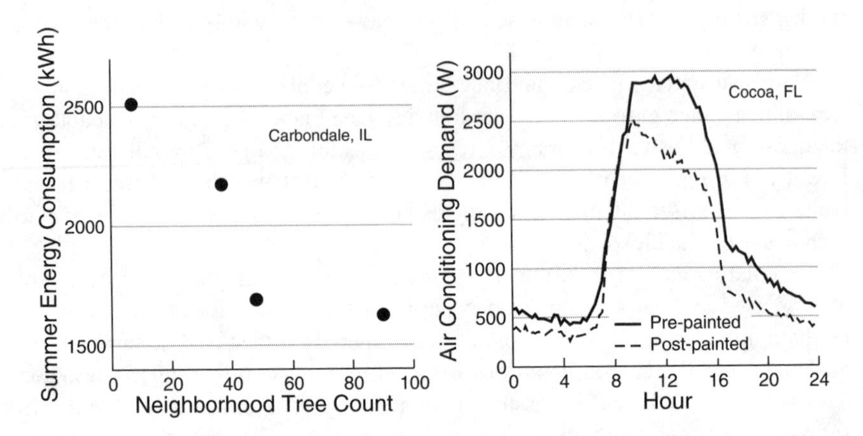

Figure 3.14: At left, shade trees reduce summer energy consumption by about 25% for houses in Carbondale, Illinois (after Carver et al. 2004). This study compared one residential area with 7–8 trees per house on 0.25 hectare lots with another having 2–3 trees per house on 0.14 hectare lots. At right, white paint on the roof of a small strip mall in Cocoa, Florida, changed reflectivity from 28.8% to 75.3%, producing a 24% energy savings (after Parker et al. 1997).

What about energy? The results on energy reduction brought about by trees, in the left plot of Figure 3.14, come from a small study of 18 single-family homes across two sites in Carbondale, Illinois, during the summer season. One site had homes "ideally" surrounded by trees and plants, and the other had nonideally vegetated homes. I've plotted the data here as energy use against tree count: more trees, lower summer energy consumption. Across these sites, trees cool houses mostly by providing shade and reducing air temperatures through transpiration and reflecting sunlight.

Is the energy reduction cost-effective? From my electric bill, I calculate the savings between these two neighborhoods at about $100 per cooling season, or somewhere around $3 saved per tree per summer (about 20 kWh). I once called a tree service to take down part of a split tree leaning toward my house. That job cost a few hundred dollars, but I know of people paying $2,500 to have a very large tree removed, avoiding what might be very high repair costs.

With regard to balancing costs and benefits for a single tree, suppose a 50-year lifetime and a cooling value over that lifetime of, perhaps, $150. If it's close to a house, the need to call a tree service to take it down erases the entire benefit and more.[64]

Alternatively, the simple act of painting roofs white also provides direct energy reductions, as shown in the right-hand plot.[65] Generally, increasing the albedo of house coatings yields energy savings of 20 to 80%.[66] It isn't as beautiful as a tree, but painting a roof white costs very little in the right environment. A 1998 study showed that changing albedo and planting trees in the Los Angeles basin cost 0.4 cent for each kWh saved.[67] One of my 2007 electric bills shows a cost of around 15 cents per kWh. Even if the cost from 10 years ago escalated a factor of ten, the savings seem really good. But keep in mind that savings from painting the world white aren't universal. Simulations of buildings in Toronto, Ontario, indicate unimpressive energy savings from cool roofs and trees, mostly because reflective roofs *cost* energy in the winter, a time when homeowners want low albedoes, letting sunshine warm up a house.[68]

Coming back to trees, we see here summer cooling energy reductions, and energy reductions mean lower fossil fuel consumption. A Los Angeles study claims that the average tree prevents 18 kg of emissions each year from power plants while sequestering 4.5 to 11 kg directly. The average tree reduces emissions in several other cities by 10 to 11 kg/year.[69] These numbers agree with the conversion between wood and electrical energy discussed in Figure 3.16.

In summary, well-planned energy-reduction strategies produce savings in the following proportions: 20% from cool roofs, 30% from shade from trees, and 37% windbreak from trees. Remaining savings come from citywide temperature reductions through broadly adopted house-scale strategies.[70]

Trees help small houses keep cool and break even for heating.

Air Conditioning

Building Type	Total cooling (GWh)		% Change with trees			Energy use change (GWh)		
	with trees	no trees	Direct Shade	Air Temp	Wind Speed	Temp/ Shade	Wind	Total
1–4 Family	438 vs.	556	13.3	11.3	-3.3	74	44	118
5+ Family	88 vs.	99	3.9	8.8	-2.0	4	7	11
Sm-Med C/I	297 vs.	314	3.1	3.0	-0.8	10	7	17
Large C/I	459 vs.	470	0.0	3.1	-0.7	0	11	11

Heating

Building Type	Total heating (TJ)		% Change with trees		Energy use change (TJ)		
	with trees	no trees	Direct Shade	Wind Speed	Temp/ Shade	Wind	Total
1–4 Family	11,367 vs.	11,416	-4.1	4.5	-464	513	49
5+ Family	1,059 vs.	1,074	-1.2	2.7	-13	29	16
Sm-Med C/I	3,032 vs.	3,048	-1.1	1.6	-33	49	16
Large C/I	4,675 vs.	4,739	0.0	1.3	0	64	64

Figure 3.15: Energy savings for cooling and heating in Sacramento, California (after Simpson 1998). Overall, winter heating uses much more energy than summer cooling. The two main effects of trees, reduced wind speeds and increased shade, have competing energy consequences for smaller residential buildings. In summer, more shade (and resulting cooler temperatures) and more wind reduce energy costs, while in winter, less shade and less wind reduce energy costs. Summertime shade benefits greatly exceed the costs of reduced breezes, while in winter, the advantages and disadvantages roughly balance. One gigawatt-hour (GWh $= 10^9$ Watt-hours) equals 3.6 terajoules (TJ $= 10^{12}$ Joules).

Trees reduce summer energy use, but what about winter? At least in the Sacramento climate, the tables in Figure 3.15 show that heating uses much more energy than cooling, about 20 times more, and trees affect that energy use very minimally. In summer and winter, trees shade buildings (making buildings cooler) and reduce wind (lessening heat transfer). More shade and more wind are good things in summer, but during the winter, energy efficiency demands less of both. Deciduous trees — those that lose their leaves — trade off the competing demands best, but not perfectly. They provide shade but reduce wind in summer, and shade less but let breezes pass through during the winter. These results show that increased winter energy costs from shade roughly offset benefits of reduced wind speed, with just a little bit of net benefit. Overall, it's estimated that Sacramento saves almost $19 million in annual energy costs due to the presence of its roughly 3 million urban trees, about $6 per tree, and about $48 per resident.[71]

Putting your house in a park, or better yet, turning your lot into a park, can have tremendous energy-saving possibilities. One study moved eight potted trees around a couple of houses in Sacramento, finding energy savings of up to 30% for cooling.[72] These savings increased to 50% for trailer houses, which tend to have lower insulation and leakier construction.[73] Another study, during warm days in Davis, California, measured 2C temperature reductions within a small forest, 150 m by 300 m, compared to bare fields. Of course, that's a big urban park full of trees, about 4.5 hectares, something over 10 acres. However, the study noted that such a big stand wasn't needed: Five meters (15 ft) into the stand the temperature reduction reaches 65% of its maximum (see the sharp thermal boundaries in Figure 2.1), with 50% windspeed reduction.[74] In this case, researchers argued that reduced wind minimizes infiltration into leaky houses, preserving the interior air-conditioned air.

The balance between winter and summer energy costs and benefits depends crucially on location, and results for Sacramento have doubtful implications for Durham, North Carolina, or Minneapolis, Minnesota. These studies demand repetition in each urban area, right down to individual neighborhoods. For example, newer, well-insulated houses have lower possibilities for tree-related energy savings.[75] Across a city, imagine having as much information as possible on buildings, trees, and impervious surfaces, along with electricity, heating, and water use: Crunch all the numbers to understand the most efficient situations to improve the least efficient.[76]

Wood has low energy content.

Tree Size (inches)	Trees per cord	Kilograms per tree	Gas equiv. (gallons)	Electricity days
5	46–55	36	4.4	4.0
6	21–33	64	7.8	7.1
7	14–18	112	14	12
8	9–14	156	19	17
9	6–9	240	29	27
10	4–6	360	44	40
16	2	900	110	100
22	1	1800	220	200

Energy Source	Wood (lbs.)
#2 Fuel oil	
one gallon	22.2
Natural gas	
one therm (100 ft^3)	14.0
Propane	
one gallon	14.6
Electricity	
one kilowatt-hour	0.59
Coal	
one pound	1.56

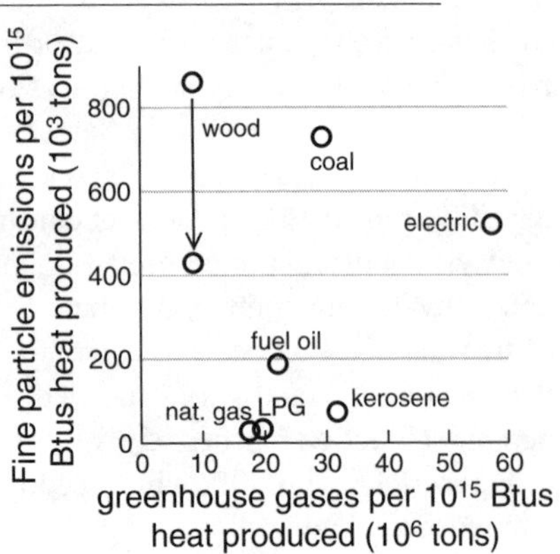

Figure 3.16: The tables compare trees and wood with personal energy use and the equivalents of various energy sources (after DeWald et al. 2005). Americans' annual per capita energy use equals about four 22-inch diameter trees. The plot compares source emissions of fine-particles and greenhouse gases for equivalent amounts yielding 10^{15} Btus of energy (after Houck et al. 1998). Wood wins in terms of greenhouse gases, but loses badly for fine particle emissions. Coal and electricity are generally worst, and natural gas has the lowest emissions. Newer EPA-certified wood-burning stoves (at the arrowhead) greatly reduce emissions.

The recent push toward biofuels takes us back to the future. Wood, a fine biofuel, provided most of our energy 100 years ago (Figure 3.1). The top table in Figure 13.16 relates tree size to energy content via the standard measure of wood, the cord, a stack of split wood 4 ft high by 4 ft wide by 8 ft long, with a mass of about 1,800 kg.[77] Over the year, the average North Carolinian uses about 1,100 kWh per month, or about 37 kWh per day, which equals about 10 kg of wood per day.[78] A wonderful University of Nebraska publication relates the energy from wood to other sources in the bottom left table.[79] My pickup truck's gas tank holds about 10 gallons. Assuming a gallon of gasoline and a gallon of fuel oil have the same energy content, filling up my gas tank equals cutting down and burning up a 6.5-inch diameter tree. Do that once a week, and I know my yard will be quickly deforested.[80]

Wood provides heat. Space heating accounts for 8% of the United States' present energy use, of which 86% comes from fossil fuels and 8% from wood.[81] How good is wood heat? On the one hand, in a wood-chipping operation the heating energy gained was almost 28 times the energy invested, with the greatest expenditure being transportation. The energy return for unchipped firewood ought to be higher.[82] On the other hand, wood is one of the worst sources of heat in terms of fine-particle emissions. In terms of greenhouse gases, CO_2 dominates the emissions for wood heating, with methane and NO_x being negligible. Remember, though, that trees contain "modern carbon," not fossil carbon, meaning the release of trees' CO_2 doesn't increase present-day concentrations. Transporting wood to the fireplace, however, counts against its carbon budget.[83]

These tables show why we changed from wood to fossil fuels (Figure 3.1): Fossil fuels contain a lot of energy in smaller volumes. Though old steam-powered trains did it, nobody wants to pull a trailer full of cordwood behind a steam-powered automobile. Can we convert this old-fashioned, low-density plant material into a new-fashioned, high-density fuel like ethanol? Unfortunately, much of the recent discussions focus on energy and nitrogen-intensive biofuels farming, and I certainly don't blame farmers for seeking greater market demand, given their economic realities (see Figure 1.7). But efficient fuel-farming requires rapid carbon sequestration by plants (in other words, they grow fast), with desired rates nearing 22 Mg/hectare/year, or 2.2 kg/m^2 of dry biomass. Seemingly, short-rotation woody crops, like silver maples, sweetgums, and poplars, have high potential with regard to biomass production.[84]

Diverted waste-stream biomass adds another fossil-carbon-free energy source,[85] an especially great idea if the biomass would otherwise go into landfills. These wastes might include corn stalks and other crop wastes, as well as paper products, but low-till, soil-enhancement practices might take a hit.

Durham citizens use more energy than local forests can provide.

Energy Budget

Energy Use

Gasoline Energy Content: 115,000 Btu/gallon
460 gallons/person/year
= 53 million Btu/person/year
Electrically supplied energy: 3,400 Btu/kWh
NC Electricity Use: 14,400 kWh/person/year
= 49 million Btu/person/year

Energy Sequestration in Trees

Energy content in Pine/Oak: 14,000 Btu/kg
Biomass fixation: 1 kg/m^2/year average
= 4,047 kg/acre/year
= 56.7 million Btu/acre/year

Energy offset

Fossil-fuel energy offset by tree fixation
(102 million Btu/person/year)
/(56.7 million Btu/acre/year)
= 1.8 acres/person/year

Area of Durham County: 186,000 acres
Population of Durham County: 225,000 people
(186,000 acres/225,000 people)
= 0.83 acre/person

Figure 3.17: A calculation of Durham County citizens' energy footprint. On a per capita basis, energy use represents 1.8 acres of net primary production by plants, but the county only has 0.83 acre on a per citizen measure. These numbers for personal use underestimate total energy use, from Figure 3.1, of 100 quadrillion Btus for 300 million Americans, yielding 330 million Btu/person/year. This energy offset would require more than 5 acres.

Here's an "energy footprint" calculation (Figure 13.17). For North Carolina, per capita energy consumption, both fuel and electricity (see Figure 3.2), sums to about 100 million Btu/year. Given Durham County's 0.83 acre/person (3,400 m^2/person), these numbers mean that North Carolinians "locally" add about 30,000 Btu/m^2/year in heat through energy use, or about 80 Btu/m^2/day. We aren't directly heating up the Earth by using energy: In comparison, solar radiation adds about 16,400 Btu/m^2/day at Earth's surface.[86] There's a lot of solar heat energy compared to our measly energy use.

How does our energy use relate to trees? Suppose we could achieve biomass fixation of 1 kg/m^2/year (see Figure 1.6), representing 14,000 Btu/m^2/year (about 0.3% of the incident light energy), close to the natural amount, and about half our localized energy use. In other words, if we could harvest all the biomass fixed yearly off of 1.8 acres, in a sustainable way assuring the same harvest year after year, that biomass would provide the energy each North Carolinian uses in one year.[87] But Durham County's residents only have 0.83 acre per person, including urban areas. Our energy footprint exceeds our county boundaries.

If we could grow energy supercrops that fix 10% of incident light energy every day of the year, we'd produce 1,360 Btu/m^2 of biomass each day.[88] If we could then convert 10% of that biomass to usable energy to power cars and dishwashers and light bulbs, we'd have 136 Btu/m^2/day, about double what North Carolinians use today.

If the calculation isn't too silly, let's get a bit more realistic about biomass conversion. Hoped-for biomass fixation for short-rotation woody crops tops out at 84 Btu/m^2/day *before* conversion to fuels.[89] With a 10% biomass-to-energy conversion efficiency, that would provide about 8 Btu/m^2/day, roughly 10% of the energy we use personally. Of course, we couldn't live and produce food crops on the land on which we're growing energy crops, nor could we have slow-growing trees, wildlife, and healthy soils.

Earlier I mentioned the terrestrial Earth has about 5.8 acres per person across the globe, with a world average biomass fixation around 1 kg/m^2/year. Harvesting all of this biomass, each human would have about 24,000 kg (and other organisms none), representing 336 million Btus each. As I calculated here, North Carolinians personally use about 102 million Btus, or about one-third our individual biomass harvest.[90] We already capture about 40% of the world's net primary productivity for many purposes, a capture that already appears beyond sustainability, as evidenced by environmental degradation and high species extinction rates.[91] Seemingly, urbanites' focus must be on using urban trees to reduce energy demands, not fulfill their energy desires.

Chapter 4

Emissions and Urban Air

Plants, animals, and machines all contribute chemicals to the air we breathe. Over hundreds of millions of years, the atmosphere changed in response to biological and geological processes, and, in turn, organisms evolved in response to a changing atmosphere. In this chapter I look at the issues involving air, covering the causes and consequences of contributions from both human and nonhuman organisms.

Humans usually dominate emissions in urban areas. Air quality studies categorize emission sources as either point or nonpoint, roughly corresponding to fixed or moving, respectively. Both types of emissions roughly match maps of population density, at least in the United States. A number of studies demonstrate the clear connection between transportation and urban air quality, including an opportunity presented by the 1996 Summer Olympics in Atlanta, Georgia. Looking on the positive side, the United States undertook serious efforts to improve air quality, with good reductions measured and reported by the Environmental Protection Agency. Still, the United States measures some of these emissions in pounds per person per day.

Vegetation and soils also emit chemicals, often as a by-product of evolution in the face of herbivory, competition, and environmental stresses that yielded chemical responses and defenses. These emissions can reduce air quality in urban areas, with some tree species being worse "violators" than others. Using various information sources, biogenic emissions across North America shows some regions with particularly high emissions, one area being the mountains of the southeastern United States, just upwind of Durham, North Carolina. Though biogenic emissions fall well below fossil-fuel emissions, considering their propensity to interact with other chemicals, they have important air quality implications.

These emissions feed into, among other pollutants, ozone formation in the hot summer months. For a more complete understanding of urban pollution, I provide an overview of the chemical reactions that link volatile organic compounds (VOCs), reactive nitrogen, sunshine, ozone, and eye-stinging pollutants. A particularly revealing example follows the emissions plume from a coal-burning electrical power plant, in this case, with emissions producing ozone just as the plume sweeps over Nashville, Tennessee. This example helps explain several urban situations where transportation produce emissions. It would be convenient to blame some regulatory agency for falling down on the job, but these ozone-producing reactions have complicated dynamics depending on the ratio of VOCs to reactive nitrogen. Sometimes one thing should be reduced, sometimes the other, and sometimes climate plays a major role. Given that these emisssions vary greatly over the United States, as we saw in the previous chapter, nationwide regulations seem elusive.

Still, cities have higher ozone levels than rural areas: in the worst of times more than double the ozone levels. Although in a later chapter we examine ozone's health effects, here I show that vegetation also suffers from high ozone levels. Indeed, high ozone levels reduce wheat and potato yields by about 30%, meaning that farms surrounding cities might have problems beyond urban sprawl.

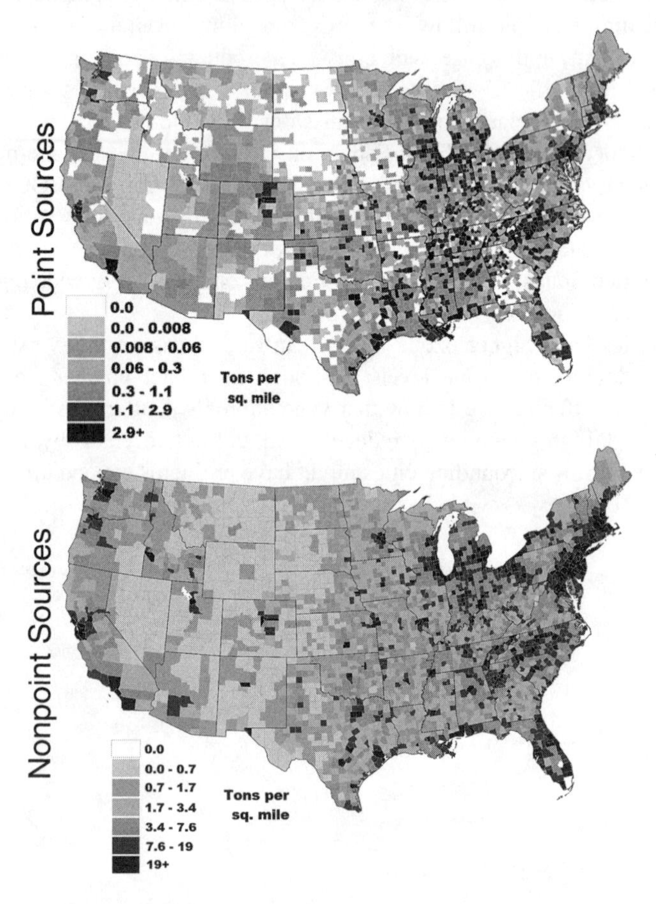

Figure 4.1: Human sources of volatile organic compounds (VOCs) distributed across the contiguous United States in 2001 (images from U.S. Environmental Protection Agency). Point sources (top image) include power plants and factories with high emissions, while nonpoint sources (bottom image) cover vehicles, dry cleaners, and the many small emission sources too numerous to inventory individually. Note the different scales on the two sources. ($1\ \text{ton/mi}^2/\text{yr} = 40\ \mu\text{g/m}^2/\text{hr}$.)

Figure 4.1 shows human-caused emissions of volatile organic compounds (VOCs) across the United States.[1] Point sources represent power plants and factories, things that stay fixed on the ground, generating emissions from a single location, and, perhaps, above some emission level.[2] Point source emissions take place close to cities in part because people want electricity and long transmission lines waste electricity. And they take place not just anywhere close. Evidence exists that these point sources are more often sited near poor and minority communities.[3]

Nonpoint sources clearly are located where people live and, more importantly, drive. These sources move around on the ground, not just cars and trucks, but also trains, planes, and tractors — off-road vehicles that add up to a lot of emissions. Take a look back at the distribution of people across the 48 states in Figure 1.1 and compare with these emissions figures. A pretty strong correlation exists because the EPA uses population densities to come up with some estimates for nonpoint sources, assuming there's an average emissions per person. One can't argue too much with that approach, given constrained budgets for air sampling and the fact that 300 million Americans use spray paint and hair care products (and the like), all included in nonpoint sources.

What compounds are we talking about here? There's a slew of chemicals, but the principal components of both urban air and car exhaust include ethane (C_2H_6), ethylene (C_2H_4), propylene (C_3H_6), propane (C_3H_8), butane (C_4H_{10}), iso-pentane (C_5H_{12}), benzene (C_6H_6), and toluene (C_7H_8).[4] All of these chemicals can be substituted for the methane (CH_4) in Figure 4.7, factoring into the production of ground-level ozone.

Another important ingredient of urban air comes from burning fossil fuels: It emits a pair of nitrogen oxide compounds, $NO+NO_2$,[5] nitric oxide and nitrogen dioxide, respectively, and these reactive forms of nitrogen, together called NO_x, cause problems in our urban air.[6] Though reactive nitrogen right near the source of its emission depresses ozone concentrations (see Figure 4.14) through the reactions outlined in Figure 4.7,[7] NO_x remains after removing the ozone, and later reactions result in downwind ozone increases during daylight hours (see Figure 4.8).[8]

We'll examine a few of these complicated emission issues and their consequences when we look at ozone production later in this chapter.

Fossil-fuel use produces many pollutants.

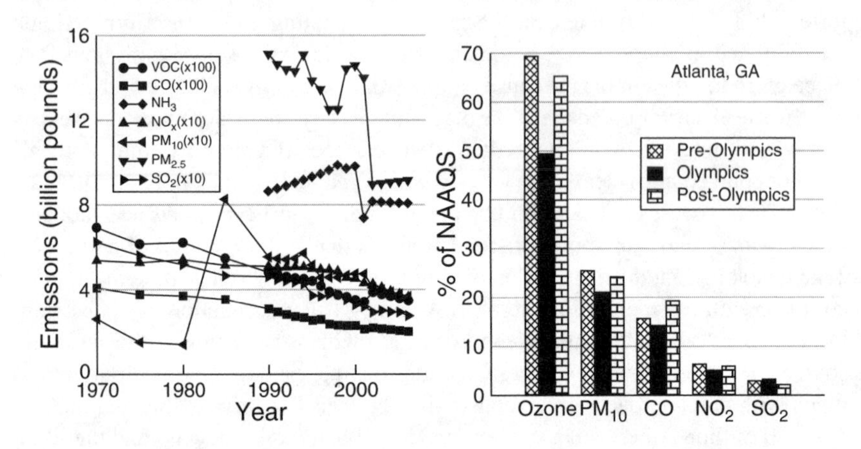

Figure 4.2: At left, annual pollution emissions from human sources show gradual decreases over the last several decades (data from the U.S. EPA). Pollutants include volatile organic compounds (VOCs), carbon monoxide (CO), ammonia (NH_3), reactive nitrogen (NO_x=NO+NO_2), particulate matter (with condensables) below 10 and 2.5 microns in size (PM_{10} and $PM_{2.5}$), and sulfur dioxide (SO_2). The right plot compares pollutant levels before, during, and after the 1996 Atlanta Summer Olympics, during which authorities reduced traffic by 22.5% (after Friedman et al. 2001). Ozone, carbon monoxide, and PM_{10} (particulate matter below 10 microns diameter) levels were reduced significantly during the Olympics. NAAQS stands for the National Ambient Air Quality Standards set by the U.S. Environmental Protection Agency.

The left graph in Figure 4.2 shows, on a national scale, that Americans put quite a bit of pollutants into the atmosphere. The EPA data separate these numbers into many source categories, sometimes humorously so, for example, with around 85% of NH_3 and PM_{10} emissions labeled "miscellaneous."[9] Notable successes lead to the decreasing trend, giving hope for further air quality improvements. Implementation of the 1970 Clean Air Act[10] reduced the various pollutants over this period, showing the power of effective legislative regulation. Most impressively, perhaps, legislation reduced VOC emissions from highway vehicles from 17 million tons in 1970 to 3.9 million tons in 2006.

Note also the change in the U.S. economy, seen in Figure 3.4, from 40% goods-producing industries in 1950 to 20% in 2000. Service industries took up the slack. Perhaps a service economy has fewer emissions than one based on producing goods: Could the emissions shifts seen here be, in part, a reflection of this economic change? One problem with this hypothesis is that transportation produces much of our NO_x and VOC emissions, and people's travels don't decrease with a service economy.[11]

One example highlights traffic's role. Atlanta imposed tight traffic restrictions during the 1996 Atlanta Summer Olympics, and the experiments, summarized in the right plot, took advantage of the situation.[12] Gasoline sales decreased by 3.9% for the month, including purchases by 1 million visitors. During the games, weekday peak morning traffic counts were 22.5% lower than usual, though total weekday traffic decreased only 2.8%, but public transportation more than doubled with 17 million additional trips!

These data represent the findings as a percentage of the National Ambient Air Quality Standards (NAAQS)[13] and clearly demonstrate that emissions affect air quality. During the games carbon monoxide and PM_{10}[14] drops were significant, and ozone levels were 28% lower. In addition, there were significant correlations between peak morning traffic counts and peak ozone concentration on each day. However, after correcting for four weather variables, temporal trends, and day of week, ozone levels were just 13% lower within Atlanta compared to 2–7% lower in neighboring areas.

What do these numbers mean on a personal level? Consider, for example, human emissions of VOCs, amounting to roughly 400 billion pounds per year. These emissions run the gamut of solvent use, highway vehicles, other fuel combustion, and so on. Dividing this huge number by the U.S. population gives a bit more than 1,000 pounds per person per year, or VOC emissions of about *3 pounds* per American per day.

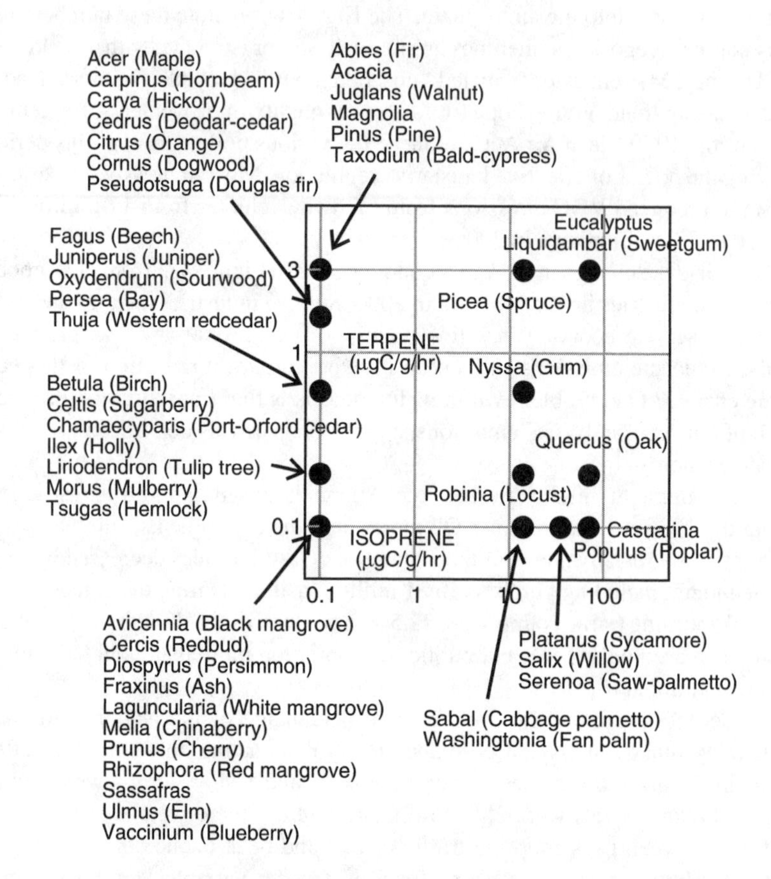

Figure 4.3: Trees release thousands of chemicals that play roles in protection from thermal stress and herbivory, attractants for pollinators, and plant–plant antagonisms. Here I plot emissions of two such biogenic volatile organic compounds, monoterpenes and isoprene, for various species (data by Guenther et al. 1994). Low-emission trees include the beautiful redbuds and super fast-growing tulip-tree, whereas the fast-growing sweetgum has high emissions. Southeastern forests support many of the native and introduced species listed here.

Natural biological processes, as well as fossil-fuel emissions, affect the chemical reactions that take place in the air we breathe. Vegetation contributes to the chemical composition of our air, and Figure 4.3 summarizes the biogenic VOC contributions from many important southeastern tree species. Even though many people fear the words "chemical" and "nuclear," solar power equals thermonuclear fusion, and an organism equals a container of chemicals. Fortunately, that massive, uncontained fusion reactor sits 150 million km away. Context is what matters.

Not to overstate Mother Nature's toxic side, but the process of natural selection produced many toxic chemicals, poision ivy and bee stings being just two personal examples. These chemicals arose as a result of natural selection, sometimes because of their role in the chemical dance taking place between plants and their herbivores, pollinator attraction, and competitive interactions with other plants. For every bit of sweet nectar that seduces insect pollinators there exists a chemical defense turning caterpillars into gooey masses. Natural selection takes place when a tree that makes a bit more of a nasty chemical deterrent to herbivores, or attractive chemical lure to mutualists, produces a few more seeds than its competing trees. Repeating this process of births and deaths over hundreds or thousands of generations, with inevitable recombination and mutations, results in evolutionary change and a diversity of chemical defenses. Hence, the biogenic VOCs over southeastern forests.

Tens of thousands of these chemicals can be called biogenic VOCs, with broader classes defined according to chemical structure.[15] I plot just two VOCs for a variety of species.[16] Terpenes,[17] a broad class of compounds of one or more isoprene molecules, may play a role in plant defense against herbivores and fungal infection, and isoprene plays a role, perhaps, in thermal protection.[18] Keep in mind that these estimates have very high levels of uncertainties,[19] and all of the emissions are normalized to 30C and average sunlight.[20]

Notice how the conifers are located high on the terpene axis and how some trees have high emissions on both axes. The tulip tree, one of my favorite fast-growing trees in the southeastern United States, has rather low emissions. Just like VOCs from burning fossil fuels, these natural VOCs sometimes play a role in ozone formation, though pretty much in any context I'd rather breathe forest air than the emissions from tailpipes.[21]

VOCs produced by trees vary across the contiguous United States.

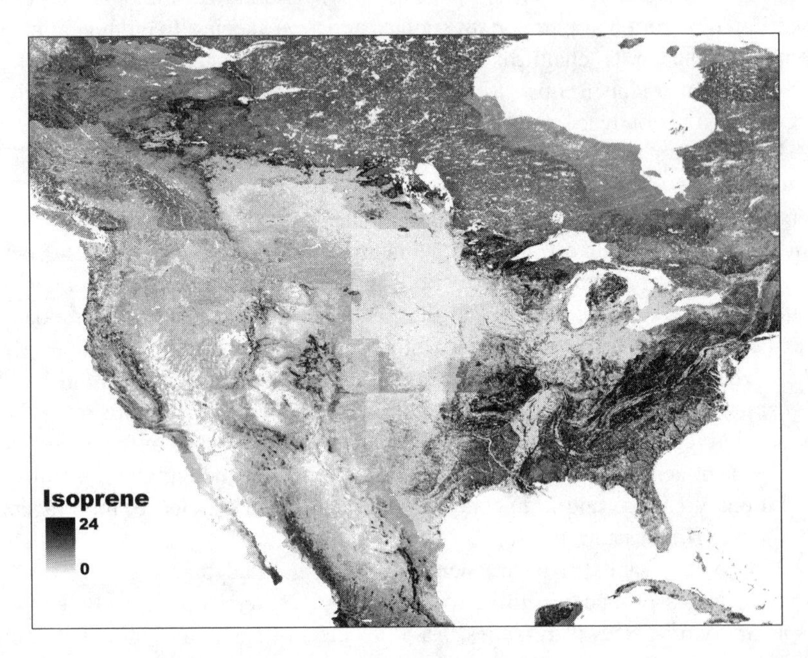

Figure 4.4: Combining trees, vegetation, and soils, this map shows the resulting isoprene emissions (μg/m^2/hr) across North America. Not surprisingly, biogenic emissions differ quite significantly from the human emissions maps of Figure 4.1, but their comparison indicates the differing importance of anthropogenic and biogenic sources of VOCs dependent on geography. Note that emissions of 1 ton per square mile per year (the unit used in Figure 4.1) equals, on average, 40 μg/m^2/hr. (Image courtesy of Alex Guenther; after Guenther et al. 2006.)

The images in Figure 4.4 map biogenic VOC production across the contiguous United States, as well as much of Canada and Mexico, made by folding together species emission data with landscape-level information about local forest composition.[22] Emission rates ranged up to 24 $\mu g/m^2/hr$[23] at standard conditions for air temperature on a clear day.[24]

Let's try keeping natural emissions in perspective: Human emissions in Figure 4.1 ranged up to 20 tons per square mile, or 800 $\mu g/m^2/hr$. Although biogenic concentrations are relatively low, I'll show in Figure 4.6 that their contribution to chemical reactions can be relatively high in some places. Overall, the U.S. Environmental Protection Agency estimates natural emissions are about double that of human sources.[25]

Vegetation can also help reduce pollution. At first glance, it seems counterintuitive that cities could use trees to reduce air pollution when forests are a part of the problem; for example, some major biogenic emissions from beautiful southeastern forests sit just upwind of Durham. But, simply put, trees don't release much reactive nitrogen, NO_x (it was hard enough for them to get the nitrogen in the first place), and NO_x plays a crucial role in the formation of hazardous urban air toxics (see Figure 4.7).[26] Pollution from burning fossil fuels and tree pollution aren't the same thing.

Vegetation in urban forests can reduce airborne pollution in two ways: by physiologically fixing pollutants into plant matter and by slowing down air within, say, a tree and letting pollutants settle out onto the ground, leaves, and branches. Imagine putting a plastic, artificial Christmas tree on New Year's Day in your backyard, under an umbrella to keep the rain and sun off but not blocking air circulation.[27] When you bring it back inside the following December, would it be clean? No, because the tree acts much like a stack of filter paper. Of course, given a choice for cleaning urban air, I would prefer a city full of evergreen trees to one full of plastic Christmas trees.[28]

Big particles like SO_4, NO_3 NH_4, and H ions, deposit on surfaces, making the discarded plastic Christmas tree yucky, whereas SO_2, HNO_3, NO_2, O_3, NH_3, can end up inside leaves.[29] In particular, ozone (O_3) harms trees, leaves, and seedlings through myriad ways once the ozone enters through the stoma, the little openings that let CO_2 and transpired water pass. It also causes visible damage to commercial fruits and vegetables that depend on their attractiveness for high value. One demonstration in Lahore, Pakistan, showed that the filtration of urban air having high-ozone levels increased wheat and rice yields 40% (see Figure 4.13).[30]

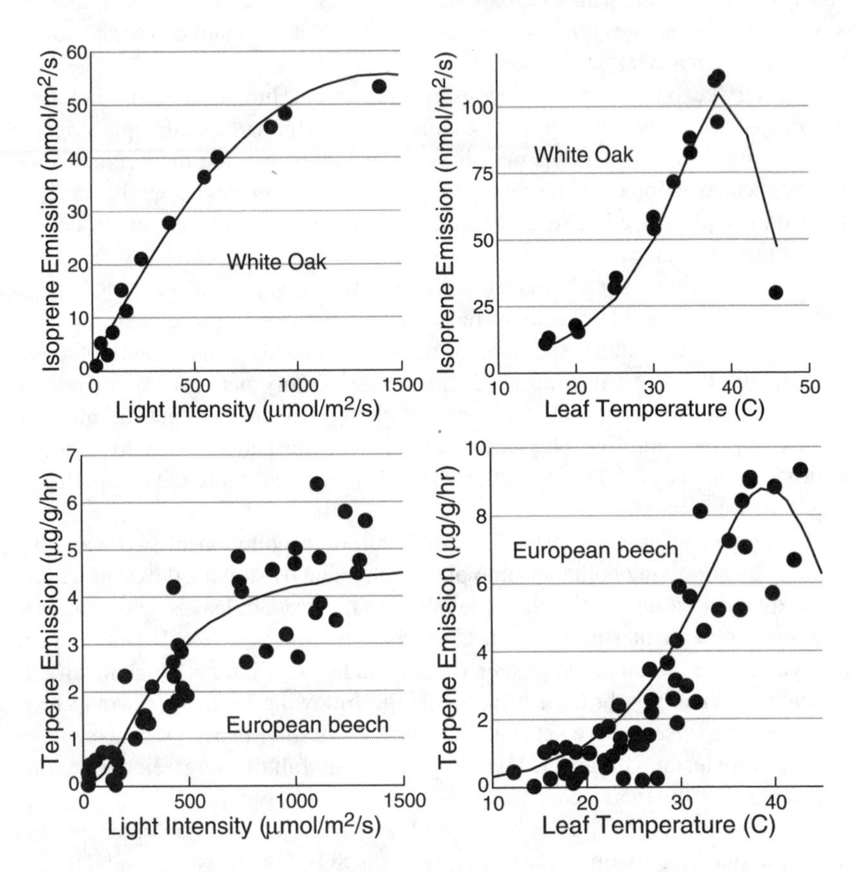

Figure 4.5: The isolated dots for tree emissions in Figure 4.3 hide the large variation actually observed. Here isoprene emissions of white oaks, measured on intact leaves and branches (after Baldocchi et al. 1995), and mono-terpene emissions of the European beech (after Dindorf et al. 2006) show great variation as a function of the light level and temperature. Hot and bright summer days yield high emissions, contributing to high VOC concentrations.

The emissions variation shown on the plots in Figure 4.5 underlines the simplistic view of species-level biogenic emissions implied by Figure 4.3's data points. Most of the measurements reported in Figure 4.3 used individual leaves, and some values resulted from just one or two measurements on a single species of the genus. These emissions estimates are difficult measurements, and many complicating features and realistic environmental conditions must be sacrificed: You just can't stick a whole tree into a plastic bag to measure its emissions. Because of these difficulties, little information exists on how these emissions change under environmental stresses like drought or polluted environmental conditions.

Using just two factors important to urban environments, light[31] and heat, these plots show how VOC emissions by beech and oak trees vary with environmental conditions.[32] The left-hand plots show that isoprene emissions saturate at high sunlight, presumably when plants need thermal protection more than ever. Emissions also depend on details like leaf age — young leaves have lower emissions — while other factors, like humidity, are less important.

One might conclude that these emissions curves mean that tree physiology changes with urban heating or global warming. When scientists speak of uncertainties about climate changes taking place alongside global warming, and say they really don't understand all of the implications, this is the kind of detail and uncertainty being discussed. How will changes in tree physiology feed back into climate change? These are hard questions, but no one should dismiss global warming because scientists don't understand details like these.

The white oak data at the top comes from measurements on individual leaves, with emissions measured with respect to leaf area, not leaf mass, making the comparison with Figure 4.3 difficult.[33] The beech data arise from measurements of a sunlit branch at the top of a 160-year-old tree, incorporating some self-shading, where one leaf shades another.[34] Figure 4.3 represents terpene emissions for beech as about 0.5 μgC/g/hr, but here the value ranges from 0 to 10 μg/g/hr,[35] showing the roughness of these average values. So, yes, trees "pollute," and some pollute more than others, but it's hardly appropriate to compare emissions from organisms with fossil-fuel emissions from the machines humans use.

VOC sources vary in place and time.

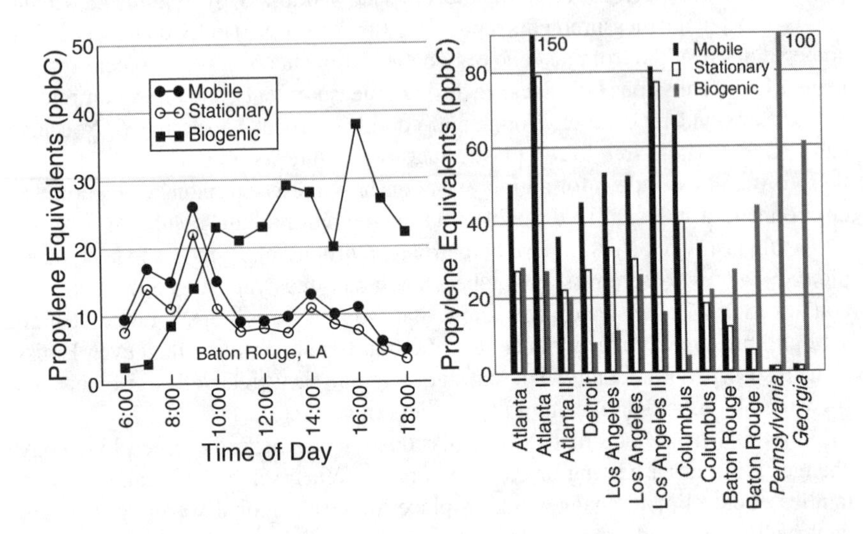

Figure 4.6: At left, measurements taken on the Louisiana State University campus in Baton Rouge, averaged over July 18–26, 1989, show the changing contributions of the three sources over the course of the day. The plot at right presents the relative reactivity contributions of VOCs from mobile, stationary, and biogenic sources (both plots after Chameides et al. 1992). Data come from samples taken in the 1980s. *Pennsylvania* and *Georgia* indicate rural sampling stations.

As a quick review, there are two broad classes of anthropogenic sources of volatile organic compounds — mobile (think vehicles) and stationary (think power plants and industrial centers) — and a multitude of biogenic sources, which include vegetation and microbial activity in soils. These different sources of VOCs change in importance throughout the day, as the left graph in Figure 4.6 shows.[36] All three sources are made relative to a common measure, the reactivity of propene, made necessary due to the high diversity and differences of chemical species of the three sources.

The early morning peak comes from rush-hour traffic[37] (see Figure 4.14), and the subsequent reduction takes place because sunlight breaks apart NO_2, as shown in Figure 4.7, which then reacts with the VOCs. We see a similar situation of rapidly depleted emissions in the power plant plume and city, as shown in Figure 4.8.

Understanding the changing importance of these various sources, with one goal being reasonable emissions regulations, requires a comparison between the many chemical species produced by vegetation, like isoprene, C_5H_8, and the vast array of anthropogenic species, the main ones being ethane, propane, iso-pentane, ethylene, acetylene, toluene, and so on.[38] Reactivity to OH serves as the basis of comparison, as all of these volatile organic compounds, including methane (CH_4) and carbon monoxide (CO), react with OH as the starting point for generating ozone and other chemical irritants. In the reactions depicted in Figure 4.7, each molecular species can replace methane along the bottom reaction chain. What we want to understand is how much each chemical contributes to, say, ozone production, taking into account its concentration and its ability to react with OH. Thus, each VOC species gets scaled into an identically reactive amount of propene (also called propylene), which has chemical formula C_3H_6.

In the right plot we see, quantitatively, the wide variation across the country in VOC source contributions that we observed visually in Figures 4.1 and 4.4. The southeastern United States has much higher biogenic emissions, and the difference in location means a greater amount of VOCs, resulting in a NO-limiting situation. Because of this difference, different things limit ozone production in forests and cities. In forests there may be lots of VOCs, but the relative absence of NO_x might limit ozone production. NO_x comes from fossil fuels, and siting coal-fired electrical power plants in rural southeastern areas certainly can introduce nitrogen emissions, with tremendous ozone implications. In cities and their surroundings, there's lots of NO_x, but ozone production tends to be limited by VOCs. In other words, ozone reduction strategies depend on these different limiting situations. Imagine the complexity of creating nationwide emissions guidelines!

VOCs, reactive nitrogen, and sunlight lead to ground-level ozone.

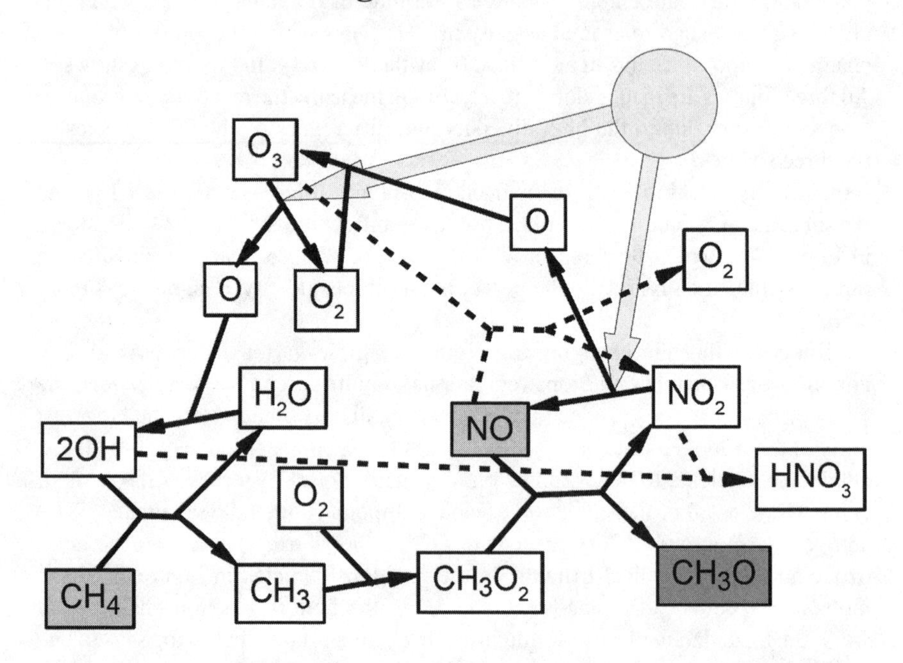

Figure 4.7: Summary of the reactions linking volatile organic compounds (VOCs), nitrogen emissions, and ozone. Light gray boxes represent VOCs, like methane, CH_4, and nitric oxide, NO (a result of burning fossil fuels). High ozone levels can arise from both types of emissions on warm, sunny days. Light splits nitrogen dioxide, NO_2, freeing an oxygen atom to react with an oxygen molecule, O_2, producing ozone, O_3. At the same time, light, as well as interactions with nitric oxide (dashed lines), degrades ozone. Many factors, including the relative concentrations of VOCs and reactive nitrogen involved in the competition for OH, determine ozone levels.

Let's put these emissions into the unifying context of atmospheric chemistry. Our atmosphere is one big beaker of chemicals, lit up and heated by the Sun, stirred by the weather, with humans adding substantial novel and bizarre pollutants, as well as breathing it in and out of their lungs. This plot caricatures ground-level ozone (O_3) production resulting from this big atmospheric stew.[39]

Ozone can be good or bad, depending on where it floats. Ozone in the stratosphere — way up above 10 km — protects the biosphere from damage by high-energy, short-wavelength sunlight. Ozone in the troposphere — way down here near the ground — does direct damage to living organisms (see Figure 4.13).

In Figure 4.7, I represent by the light gray squares both emissions by people burning fossil fuels and naturally occurring VOCs, and final products by the dark gray square. My example uses methane, CH_4, and other chemical names listed include O: oxygen; O_2: oxygen, or diatomic oxygen; H_2O: water; OH: hydroxyl molecule; NO: nitric oxide; NO_2: nitrogen dioxide; HNO_3: nitric acid; and CH_3O: the methoxy radical. Along with the ground-level ozone, these final products cause a variety of discomforts.[40] I'll focus quite heavily on asthma and heart problems in the next chapter.

Of course, my diagram greatly simplifies the picture, yet it provides the essentials for the relevant depth of understanding needed to help interpret other results presented in this chapter.[41] Where to start? Oxygen, O_2, gets split by high-energy photons, light with a short wavelength less than 240 nm,[42] into two oxygen atoms; each one can then react with another O_2 molecule to form ozone, O_3. Ozone can cleave back into O_2 and an O with light around 240–320 nm being the cleaver.[43] In the early morning, most of the nitrogen exists as NO_2, with ozone removed during the nighttime by the reactions with NO, represented by the dashed lines. Once sunlight starts acting on the system, however, photochemical reactions make OH and NO available, which subsequently promotes ozone production.

Although I use methane as the input VOC, any old VOC will do, including the biogenically produced isoprene, C_5H_8. Indeed, the hydroxyl ion, OH, reacts thousands of times faster with isoprene molecules than with either CO or methane, meaning that high levels of isoprene enhance ozone formation compared with these other molecules. In this way, biogenic VOCs can have large impacts on ozone formation even when NO from fossil fuel emissions are abundant.

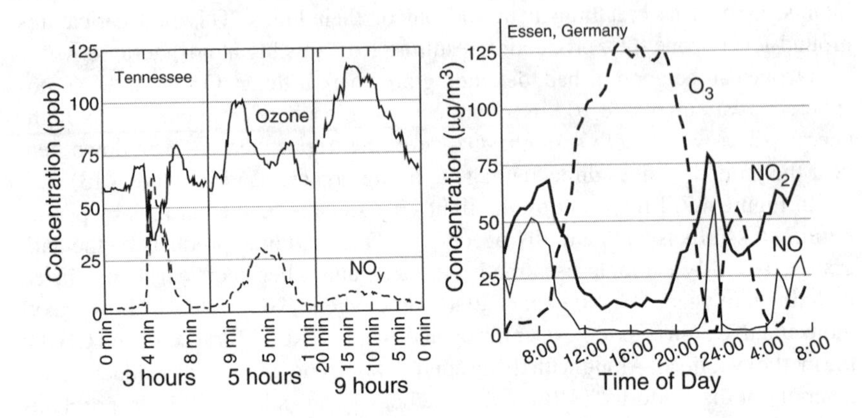

Figure 4.8: Coal-burning power plants cause downwind ozone problems, exemplified here, at left, by emissions upwind of Nashville, Tennessee (data courtesy of Noor Gillani; after Gillani and Wu 2003). Emissions occurred midmorning July 16, 1995, and the curves show ozone and nitrogen concentrations at three approximate later times. Horizontal axes display relative times of measurements from a helicopter traveling about 3–4 km per minute through the plumes. After 9 hours, the plume had drifted about 110 km downwind and expanded to about 50 km wide from its point source emission. (NO_y means total reactive nitrogen.) At right, a study from Essen, Germany, shows how morning rush-hour emissions of reactive nitrogen lead to high-ozone levels during a summer day, crashing with afternoon rush-hour and sunset. An urban heat island instability induces an atmospheric turnover, leading to a second ozone peak (after Kuttler and Strassburger 1999).

Nitrogen emisssions from coal-fired power plant smokestacks demonstrate how upwind emissions produce downwind ozone problems.[44] The graph at left in Figure 4.8 shows the changes occurring in the plume as it drifts downwind from a power plant, measured by a helicopter making successive passes through the plume.[45]

Immediately after emission, ozone levels within the plume quickly fall because the plume has lots of total reactive nitrogen, NO_y, and the NO–O_3 reaction in Figure 4.7 removes the ozone, greatly increasing NO_2. As the plume ages, several hours later and generally some 50–200 km downwind, the high levels of NO_2 become a problem because sunlight splits off an oxygen atom, which interacts with O_2 to produce ozone. This plume in Tennessee sweeps over Nashville six or so hours after emission, with ozone levels of about 110 ppb instead of the surrounding 50 ppb levels.

The graph at right, in essence, demonstrates in-place plume development. These results, from Essen, Germany, come about over the course of a day. Morning rush-hour traffic puts a large amount of reactive nitrogen into the city's air, and with the onset of sunrise, these emissions lead to high levels of ozone several hours later. With afternoon rush-hour traffic, and sundown, this ozone disappears as it reacts with newly emitted nitrogen.[46] An interesting secondary ozone peak erupts, presumably because of an atmospheric turnover induced by heat-island-caused instability. This turnover replaces the ozone-depleted air with ozone-rich air.

Results like these were important reasons for the Clean Air Act of the 1970s, when smokestack emissions produced even worse air quality problems. Ozone levels declined quite a bit after 1985,[47] with the EPA estimating that NO_x and VOCs declined 10 and 14%, respectively, in just the five years between 1996 and 2001 (see Figure 4.2). Air quality today, and in the future, may not be as grim as depicted here, but certainly needs constant monitoring. These emissions changes may, in part, reflect a change in the U.S. economy from one of production to service,[48] though Americans use more electricity and gasoline independent of this economic change. Probably the reduction arises from many factors, but it may be difficult to determine what fraction is due to which factor. In contrast to a rosy picture, some studies predict an increase in ozone problems with the increased temperatures expected with global climate change.[49] For example, at higher air temperatures, reactions might take place faster, with ozone produced closer to emission sources.[50]

Ozone production and levels have a complicated emissions dependence.

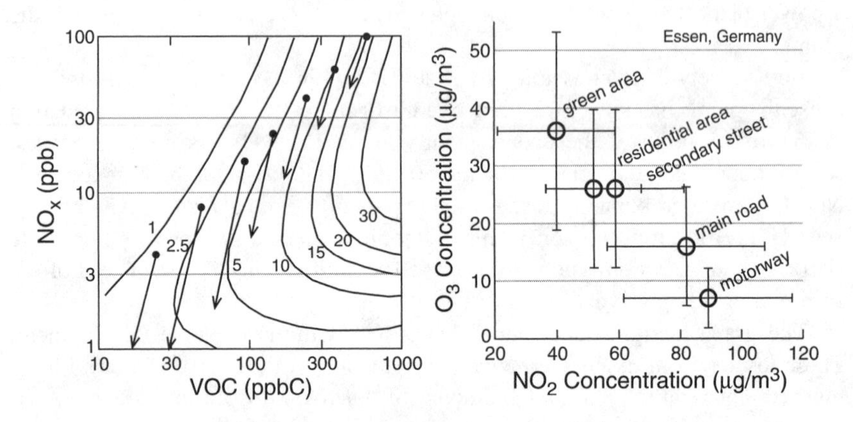

Figure 4.9: At left, the plot shows how ozone production changes with VOC and NO_x concentrations, the two important emissions ingredients in the reactions of Figure 4.7. Lines of equal ozone production, with rates of 1, 2.5, 5, ..., 30 ppb/hr, form curves on this logarithmic plot of emissions concentrations (after Sillman 1999). As the reactions produce ozone during an 8-hour sunlit day, the arrows depict how the VOC and NO_x concentrations would change from the dot to the arrow's head. Below the "ridge-line," increasing NO_x concentration increases ozone production, and increasing VOC concentrations increases it above the line. At right, data from Essen, Germany, show that ozone production variations lead to fall and winter ozone levels that differ greatly in diverse city areas, with greener areas having counterintuitively higher ozone levels (after Kuttler and Strassburger 1999).

The graph at left in Figure 4.9 summarizes complicated ozone production dynamics in an amazingly efficient and content-rich way, also making it a challenge to understand.[51] Specifically, it shows how ozone production depends on the concentrations of two important pollutants — VOCs and NO_x — defined earlier in conjunction with Figure 4.2. Solid lines that curve through the plot indicate ozone production (in units of parts per billion per hour), and the arrows indicate the direction the pollutant concentrations change over an 8-hour, sunlit period from 9 AM to 5 PM.

So what does it all mean? Here is a detailed description. Take note of the ridge-line that roughly follows the ozone production numbers 30 down to 10 and then toward the lower left corner, as if this plot were a topographic map. Reactive nitrogen and VOCs compete for OH, and the relative amounts of each determine which reaction dominates the competition (see Figure 4.7). Typically, a ratio of VOCs to reactive nitrogen of 5.5 to one balances the reactions of the two groups with OH, though that number depends on the VOC species involved. Above this ridge-line, increasing levels of NO_x reduce ozone formation because NO_2 pulls out OH, slowing down the VOC-fueled process that converts NO to NO_2. However, also above this ridge-line, increasing levels of VOCs increase ozone formation because the reaction between the VOC (methane in Figure 4.7) and NO is limited by VOC concentration. Below the ridge-line, the scarcity of NO_x limits the reaction between the VOC and NO, and variation in the VOC concentration has relatively little effect on ozone formation. At the same time, increasing levels of NO_x directly increases ozone formation.

The graph at right shows what the details mean. These results, from Essen, Germany, show the cool weather variation of ozone and NO_2 within a city, again demonstrating how reactive nitrogen depletes ozone.[52] In contrast to the actual mechanisms, the graph gives the appearance that green areas might be responsible for high-ozone levels. Of course, that's not correct, so don't go cut down all your city trees.

Suppose you're the one tasked with writing nationwide regulations for automakers regarding emissions. These complexities demonstrate the difficulties of controlling ozone levels by regulating these two emissions. How do you deal with such details as whether or not there's a power plant upwind affecting ozone formation in a downwind city, and if so, NO_x emissions in cars may either increase or decrease ozone formation? Do you worry whether the car operates in a city with high- or low-biogenic VOCs?

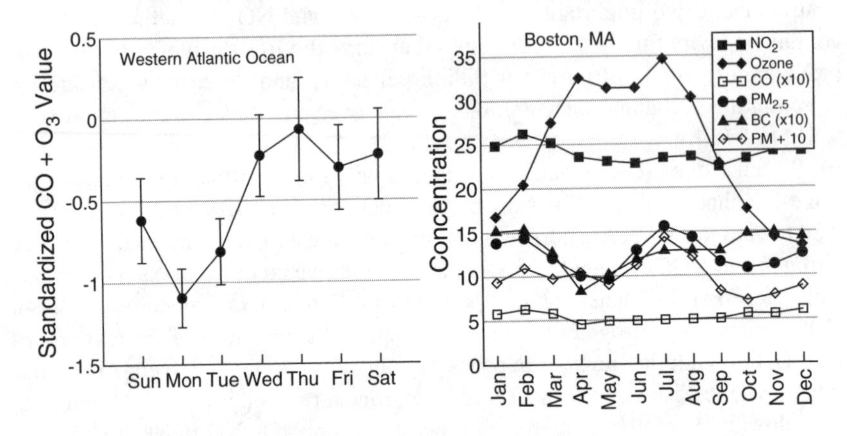

Figure 4.10: The left plot shows combined ozone and carbon monoxide concentrations over the eastern Atlantic Ocean for each day of the week averaged over the years 1991 to 1995 (after Cerveny and Balling 1998). The dip occurs from reduced weekend traffic. The plot on the right shows natural, seasonal variation in several pollutants in Boston, Massachusetts (after Zanobetti and Schwartz 2006). Numbers measure concentrations in units of $\mu g/m^3$ for black carbon (BC), non-traffic particulate matter (PM) and smaller than 2.5 microns ($PM_{2.5}$); parts per billion (ppb) for nitrogen dioxide (NO_2) and ozone (O_3); and parts per million (ppm) for carbon monoxide (CO). Ozone peaks in the summer heat.

If there's any place that should be only weakly affected by people, it might be way out in the Atlantic Ocean some 180 km off the coast of Nova Scotia, at Sable Island, about the same distance traveled by the power plant plume in Figure 4.8.[53] Unfortunately, both situations show high-ozone levels a couple hundred kilometers downwind from their pollution source. The left-hand plot in Figure 4.10 shows a combined, "standardized" measurement of carbon monoxide (CO) and ozone (O_3) at Sable Island, with strong weekly variation. No natural processes take place on a weekly basis: Something about people and their busy Monday through Friday schedule — most likely traffic — produces elevated pollution levels Wednesday through Saturday out in the Atlantic Ocean. Low levels Sunday through Tuesday rebuild from output on Monday and Tuesday, leading to high levels starting on Wednesday. In some ways, this result replicates the Atlanta Summer Olympics results in Figure 4.2 that showed decreased levels of pollutants when inner-city traffic was halted.

Here the composite variable's actual value doesn't have much meaning and should have an average of zero, but some data pairs are missing, producing an average negative value. To make comparisons with other plots, the original data set shows ozone levels mostly between 20 and 40 ppb, sometimes ranging up to 80 ppb, and carbon monoxide levels between 80 and 150 ppb, ranging up to 250 ppb.[54] The observed weekly variation, with the only possible explanation being people, is the important part.

The right-hand graph plots the concentrations of various pollutants in Boston, Massachusetts, throughout the year.[55] Some pollutants have an essentially stable level, others have variations of around 50%, but in stark contrast, ozone peaks strongly in the spring and summer months, two to three times higher than in the winter months. Resulting health problems are worse then, too (see the ozone–health study in Figure 6.2).

Two of these pollutants are just small particles of junk, $PM_{2.5}$ (particulate matter smaller than 2.5 microns — 2.5×10^{-6} meters or less), and black carbon (also known as carbon black). The pollutant listed as PM (particulate matter) comes from an estimate of nontraffic sources.[56] When you see a bus or diesel engine emitting black smoke under a heavy load, you're seeing black carbon. Diesel black carbon emissions have a size right around 50 nanometers, or 50 times 10^{-9} meters, or $1/20^{th}$ of a micron. For scale, human hair varies from 20 to 200 microns in diameter. A wavelength of visible light is about 500 nanometers, and black carbon looks sooty because it has a size that absorbs and scatters light very well. Very small stuff indeed that can get into the deepest parts of lungs.

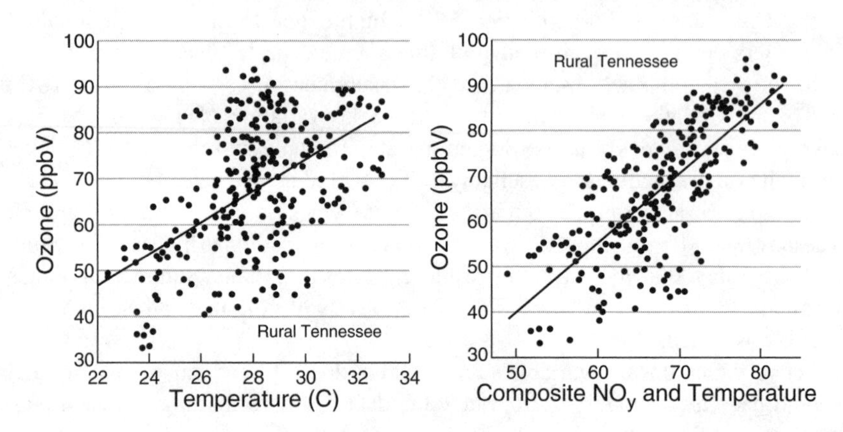

Figure 4.11: These plots depict ground-level ozone concentrations in rural areas of Tennessee (Giles County), which clearly increase with temperature, as seen in the plot at the left. However, in the other plot, data positioned closer to the regression line show that reactive nitrogen, NO_y, also plays a crucial role, as seen in Figure 4.7. The units on both vertical axes are parts per billion by volume (ppbV) (after Olszyna et al. 1997).

Ozone occurs naturally, both high in the stratosphere where it removes high-energy photons that would otherwise damage living tissues, and down near the ground in the troposphere. Data from 1991, in the left plot of Figure 4.11, show variation in ground-level ozone concentrations with temperature in the rural farmland of Giles County, Tennessee, about 75 miles (120 km) south of Nashville, Tennessee. The nearest towns are about 15 km away from the measurement site, a lightly traveled road sits 200 m away, and a two-lane highway 6 km away. Results depend on a number of factors, including cloud coverage, which changes the light available for some chemical reactions (see Figure 4.7), and atmospheric stability or stagnation (which affects mixing rates). VOC emissions from vegetation and soils also influence the data.[57]

Even better than temperature alone, a combined measure of temperature and total reactive nitrogen, $NO_y(=NO+NO_2+HNO_3+N_2O_5...)$, well predicts ozone concentration, shown at right. Rural areas have three main sources of reactive nitrogen: industry, motor vehicles, and soils. Lightning also produces reactive nitrogen, roughly one-fourth that released by burning fossil fuels, but just one-twentieth of the nitrogen made available by the synthetic Haber–Bosch process.[58] Which source is responsible? Industrial emissions come with sulfur dioxide, SO_2, and motor vehicles come with carbon monoxide, CO. Soils do not produce these additional compounds, meaning SO_2 and CO can be used to disentangle the various sources. In this study, scientists reported that when SO_2 and CO were at their estimated background levels, NO_y concentrations were about 2.3 ppbV, roughly double that in Alabama's nearby forested sites. Agricultural fertilizers explain this difference. Beyond the nitrogen accounted for by agriculture, which amounts to a rough doubling, two-thirds came from industrial sources and one-third from motor vehicles. Thus, rural areas experience agricultural, urban, and industrial pollution.

Chemical reactions generally speed up with temperature, but because temperature affects every reaction leading to ozone, one way or another, no one really knows which of the many reactions involved with ground-level ozone and emissions have the most impact. Throw in an arbitrary number of VOC species, and understanding what takes place becomes a field of detailed study. Ozone levels will likely get worse as global temperatures increase,[59] a fact that we'll discuss in upcoming plots.

Going into the plots on the next page, take note here of rural ozone levels: roughly 40–90 ppbV.[60] Let's take a deeper look at urban environments.

When it's hot, urban ozone levels exceed regulatory allowances.

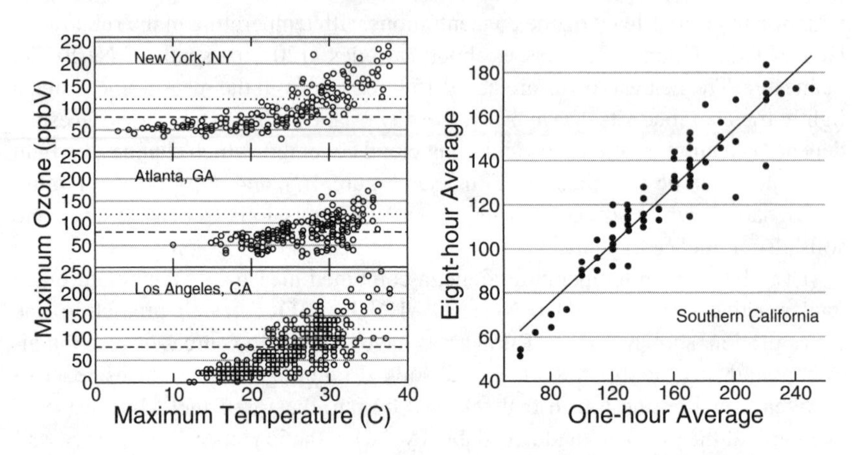

Figure 4.12: Maximum daily ozone levels increase with daily maximum temperature, as shown at left with data from Atlanta, Georgia, and New York City (data collected May–October 1988–1990, after Bernard et al. 2001), and Los Angeles, California (data collected every day in 1985, after Pomerantz et al. 1999). The dotted line represents the 120 ppb one-hour averaging regulatory standard in place before the 1997 80 ppb 8–hour standards, shown as the dashed line. The right-hand plot compares 1-hour and 8-hour ozone averages measured in Southern California during June to September 1993 (after Chock et al. 1999). The data indicate that ozone levels measured as 120 ppb 1-hour averages correspond to those measured as about 100 ppb 8–hour averages.

The graph at left in Figure 4.12 demonstrates the dependence between maximum daily ozone levels and maximum daily temperature in several cities, here showing 1988–1990 data for Atlanta, Georgia, and New York City,[61] and for Los Angeles, California, every day in 1985. Looking back at Figure 4.11, ozone levels in rural Tennessee in 1991, even in the heat of summer, the highest rural levels topped out at 90 ppbV; in these three cities the highest ozone levels are around 150–200 ppbV. Ozone levels in cities are much higher.

Granted, ozone levels fluctuate, in part, beyond human control due to weather variation, but also within human control of fossil-fuel emissions (see Figure 4.8). A moving-average method allows occasional short bursts of high-ozone levels by averaging them with background levels, thus reducing measured variability. Horizontal lines indicate the Environmental Protection Agency's (EPA) nonattainment ozone levels, dashed being the post-1997 8-hour level, and dotted the pre-1997 one-hour level. Effective May 27, 2007, the EPA set the allowable levels to 75 ppb measured as an 8-hour average. There are 24 possible 8-hour moving averages in a given day; the day is given the highest of these possible values. A city violates the standard if, averaged over the prior three years, the day with the fourth highest average exceeds this level.[62]

As mentioned before, though, complicated atmospheric chemistry underlies setting air quality standards, and the root cause of poor urban air comes from burning fossil fuels in automobiles. Imagine the statistical analyses and atmospheric chemistry that had to have gone into setting these 8-hour ozone standards: Why not the third highest average or the fifth highest average over the prior two years or four years? Those are detailed scientific (and political) issues well beyond this synthesis, yet we know that ozone differs between green and urban areas. Presumably trees (or the lack of fossil-fuel-burning cars in expansive green areas) might help reduce urban ozone problems.

Statistical measures of distributions also connect with one another, and so they do for ozone standards. The plot at right compares 1- and 8-hour averaging in Southern California. Similarly, ranges cited for 1-hour and 8-hour ozone measurements in Atlanta show, respectively, 15–129 ppb vs. 13–108 ppb in Fulton County and 18–163 vs. 15–125 ppb in DeKalb County.[63] Average values in the two standards can be set to comparable values; the real question becomes how the two measures deal with the exceedingly rare high ozone days. Similar studies carried out ozone research in Atlanta, finding that the really bad ozone days result from specific weather patterns that "cook," and recook, the same air mass over several days.[64] Bad weather events like these result in ozone violations no matter how they're averaged, and garner the greatest concern by regulators.

High ozone levels harm vegetation.

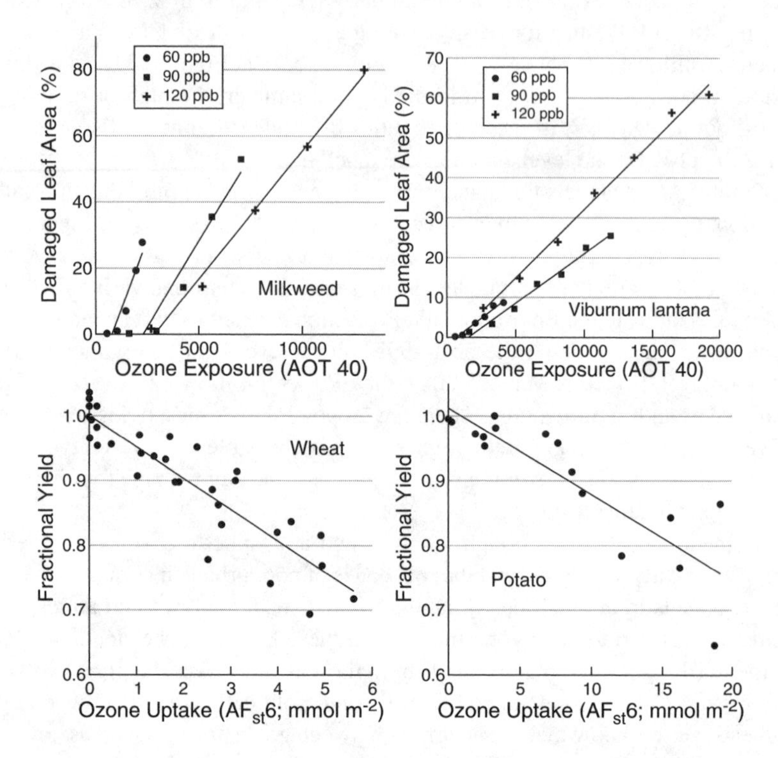

Figure 4.13: Top two plots show the fraction of damaged leaf area on two plant species, milkweed and an introduced European shrub, *Viburnum lantana,* at three different ozone exposures lasting 7 hours per day, 5 days per week, from June through August. The average exposure measure includes only values above a threshold value 40 ppb (AOT40) (after Orendovici et al. 2003). These species were two of the most sensitive of 40 species examined. The bottom plots show reduced yields of wheat and potatoes with increased ozone uptake through the stomata. Measuring ozone uptake is a more complicated — and more relevant — measure for understanding ozone damage (after Pleijel et al. 2007).

I discuss human health implications of ozone in the next chapter, but ozone hurts vegetation, too. The plots in Figure 4.13 show data from a perennial milkweed, an introduced shrub, and two important crop plants.[65] Incredibly high amounts of leaf damage and 20–30% reductions in crop yields occur at ozone levels typically found in cities — levels sometimes exported downwind of cities and power plants feeding urban energy demands.

Ozone exposure measures for vegetation use a threshold ozone level, primarily relating to nighttime ozone levels when stomata are closed. Including those time periods by averaging over these subthreshold values obscures the mechanistically relevant ozone levels. Ozone entering the stomata provides an even better measure than exposure, but a much more difficult one. It's better, in part, because high-ozone levels take place in the sunny afternoon heat when plants close their stoma to conserve water. Regardless of how it's measured, ozone damages plants.

These nonattainment ozone regulations relate to average levels, not instantaneous ones. These averaged levels are like taking a 2-hour drive and during the first hour driving 100 miles per hour (mph) and during the second hour driving 20 mph, giving an average of 60 mph. If the speed limit is 65 mph, then all is well with the averaged speed, despite the instantaneous violation during the first hour.

How harsh should we be on the EPA's allowance for variability? Clearly, averaging needs to be done with respect to nonattainment due to all the causes of ozone variability for which a city can't really be held responsible. An extraordinarily high heat wave passing by one year in three gets averaged out of this nonattainment designation. Similarly, the emissions from an upwind wildfire shouldn't produce penalties against a city. Yet, if climate change alters weather patterns over a decades-long time scale, then the definition of nonattainment (see the Figure 4.12 discussion) provides an appropriate change. An 8-hour average seems acceptable, but the devil's in the variability.[66]

The counties around Durham constitute a noncompliance area. Of course, the forests of the southeastern United States provide a little cover for noncompliance because of their biogenic VOC levels (Figure 4.4), but, then again, North Carolina counties with the forests but without the urban areas are in compliance.[67] It's hard to blame nature for the noncompliance.

Air pollution varies greatly in space and time.

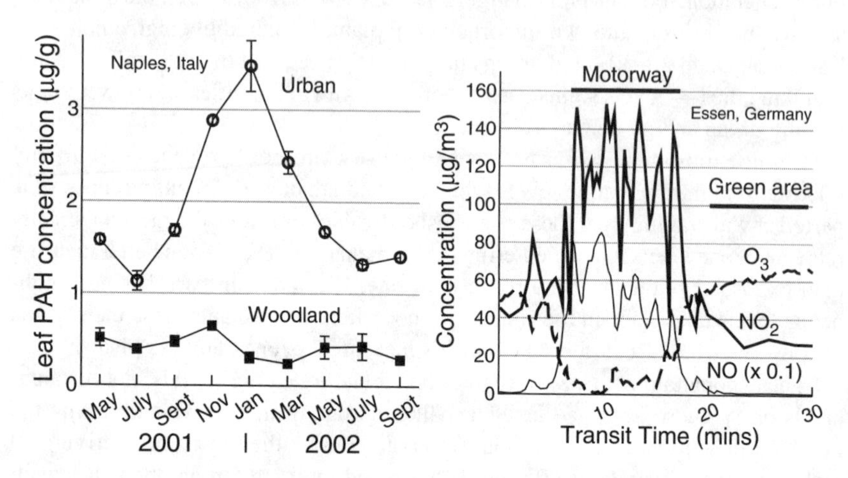

Figure 4.14: Tree leaves take up pollutants, including, in the case at left, polycyclic aromatic hydrocarbons (PAHs) in Naples, Italy (after Alfani et al. 2005). These carcinogenic chemicals look like small honeycombs, with each cell being a ring of carbon atoms. At right, a driven transect through the city of Essen, Germany, shows that trees provide better air, or that roadways provide bad air (after Kuttler and Strassburger 1999). Trees in the center of green areas likely absorb less pollution.

As discussed earlier, vegetation can reduce air pollution (see Figure 4.4). This ability arises through a variety of mechanisms, and I show one example of pollution uptake by leaves at left in Figure 4.14 In this case the pollution is a class of carcinogenic chemicals that gets absorbed into leaves' waxy layers as well as taken in through their open stomata. These studies primarily used leaves to assess pollution levels, rather than examine vegetation as a pollution mitigation strategy.[68]

Suppose a community needed to mitigate air pollution. One way to value trees from this perspective would be to calculate air pollution mitigation costs using something else, calculate pollution reduction by trees, and then determine the value of that reduction.[69] The resulting benefit-cost ratio can be anywhere from 2.2 (the benefit exceeds twice the cost) to -0.8 (trees make things worse).[70] As with any economically viable operation, the benefit-cost ratio must be greater than one (benefits exceed costs) to make trees pay for themselves as pollution control agents. Clearly, trees don't have magical pollution sequestration powers, as even discarded plastic Christmas trees can sequester pollutants by their very structure. Furthermore, trees living in parks experience lower nitrogen pollution levels, as the right plot shows, because people don't drive there: Roadside trees are the ones that have real pollution-control jobs.[71]

Of course, pollution control represents just one expected benefit, for example, pollutant-sequestering trees might be part of a small park providing recreational benefits as well as insect-controlling bird habitat. More recent detailed valuations including all of the ecosystem services provided by trees for five U.S. cities found that annual costs per tree ranged from $13 to $65, and the benefits ranged from $31 to $89 per tree, giving benefit-to-cost ratios ranging between 1.4 to 3.1.[72] Directly relevant to the Piedmont region of North Carolina, where Durham sits, one study calculated the costs and benefits of four common tree species: dogwood; Southern magnolia; red maple; and loblolly pine.[73] For small to large trees the average annual benefits ranged from roughly $31 to $112, and the costs from roughly $16 to $27. These numbers make trees seem like a great value.

However, an independent estimate of the utility that tree plantings provide in mitigating atmospheric particulate matter comes from two areas of the U.K. In one area, the West Midlands, the study concluded that quadrupling tree cover from 3.7 to 16.5% would reduce pollution levels by just 10%. Likewise, doubling Glasgow's tree cover from 3.7 to 8% would reduce PM_{10} levels by 2%. These reductions are unimpressive, though better than not planting trees, but also implying that emissions control might be the best strategy for pollution control.[74]

Chapter 5

Social Aspects of Urban Nature

People subject urban nature to the concept of "value," a concept completely irrelevant to the ecological and evolutionary context of natural ecosystems. In this chapter I first explore what it means to value something, then examine how people value vegetation, and finally investigate whether people gain social and psychological benefits from vegetation.

Certainly the development of an agrarian lifestyle subjected various plants and animals to artificial selection through their value as food and resources. A recent effort finds environmentalists pushing the idea of valuing nature for the economic benefits it provides humans through "ecosystem services." Pursuit of itemizing ecosystem services primarily involves the economics and values of marketable goods that can be bought and sold. This market valuation gained recent favor as a response to lawsuits involving environmental damage, but also as a result of several presidential "directives" that required the analysis of costs and benefits when instituting new environmental regulations.

Urban nature also gets shaped by values of another sort, so-called nonmarket goods, things that can't be easily packaged and sold, such as the enjoyment of a walk through the woods. These nonmarket influences go back in time more than 800 years, when England's King Edward I wrote laws commanding roadside vegetation be pruned to remove any hiding places for robbers. These attitudes associating vegetation and crime persist institutionally and in people's minds.

I show that, indeed, people enjoy various aspects of nature and make clear their views that life without vegetation means a lower quality of life. However, people's preferences tend toward the neat and tidy parklike setting reminiscent of new suburban lawns rather than the messy, chaotic natural forests preferred by naturalists and found perfectly acceptable by natural selection. Yet, even these stated appreciations of nature seem shallow when people flock toward unnatural, human-made environments like shopping malls.

Having views of and access to nature bring clear benefits. Public housing apartment buildings in Chicago, Illinois — housing complexes with high concentrations of economically disadvantaged people — served as the site of several studies linking crime and emotional development with vegetation. In contrast to preexisting historical attitudes, these studies indicate that places with more vegetation reduce crime, in part, by serving as social gathering points that enhance vigilance by watchful citizens. Studies of children also show that girls earn better scores on several developmental measures when they have better views of nature from their apartments. Nature also provides calming influences and helps fight childhood obesity beyond public housing situations, even extending to greater benefits with increased species richness.

Adding more urban trees provides additional "nature" beyond just a greater abundance of vegetation, including a more diverse collection of bird and plant species. Despite difficulties in measuring the value of physical and emotional health, it seems safe to conclude that urban nature makes people healthier.

What's the value of Chickpea?

Figure 5.1: A chicken barn with chickens valued at $2.60 per bird (USDA Photo by Larry Rana), chicken as a market good with a value of 30 cents per nugget, and my family's pet chicken, Chickpea, with infinite value to some family members. In one case, the chicken serves as a market good, and in the other, as an egg-producing pet. Similarly, privately owned pulp-producing forest trees only possess market value, but city trees can create positive feelings in citizens, thereby holding hard-to-measure nonmarket values.

What a difference a name makes. According to the USDA's National Agricultural Statistics Services, about 453 million chickens lived in the United States in December 2006, each worth about $2.60. Many of these chickens ended up as value-added chicken nuggets, but Chickpea (Figure 5.1), one of my family's backyard chickens, has a name. Even though she spends much of her time brooding, she's a rather happy chicken that my children love, and she recently served as a surrogate mom. Now getting on a bit in age, and really quite small, she wouldn't make much of a meal. And, of course, I'd have some very unhappy children (and spouse) if Chickpea ever ended up on the table. My family highly values Chickpea as something like a pet, making its possession of a name a matter of life or death.[1]

These feelings for Chickpea represent nonmarket goods. True, when laying, she provides an egg a day or two, making her a pet with benefits, something like two or three dollars for very free-range eggs every couple of weeks. Yet her life remains secure even if she stopped laying eggs. Having kids learn to take care of animals also has value, but it's not a market good either.

As with all market goods, we can readily understand why a grocer might get upset if a customer comes in and eats a handful of grapes, an apple, then finishes the tasting off with some orange juice, and walks out of the store without paying. These stolen items are goods paid for and owned by the grocer. Now imagine driving out in the country enjoying the views of pastures with horses, farmsteads with barns, and forested land. The farmer or rancher owns it, pays taxes, and works hard to keep it all in shape, but the travelers enjoy a free passive value in experiencing and enjoying the viewshed. Certainly, the landowner loses no market value from these passive uses. However, city dwellers and suburbanites get upset when they're denied this nonmarket good they've taken for granted because the 70-year-old landowner decides to retire (perhaps because of a debilitating illness) and cash out the market value of the property for development. Who has rights to the *view* of a privately owned property, and should they pay for that view?[2]

Building personal emotional connections between a person or a group of people and the environment can help with its protection, for example, Adopt-a-Stream, Park, Garden, or tree-planting and gardening programs in urban areas. These protective connections, like that brought about by my family's feelings toward Chickpea and her flock-mates, reflect a value determined by a process called contingent valuation. There are several fine introductions to the contingent valuation of nonmarket goods, covering both pros and cons.[3] Conserving land for wildlife relies on these concepts to make arguments against converting land into economically viable housing developments, landfills, or stripmines.[4]

S.A. Forbes (1880) estimates the value of birds.

The careful estimates of three ornithologists and experienced collectors give, as an average of the whole bird-life of Illinois [except the swimmers], three birds per acre during the six summer months. It is my own opinion that about two-thirds of the food of birds consists of insects, and that this insect food will average, at the lowest reasonable estimate, twenty insects or insects' eggs per day for each individual of these two-thirds, giving a total for the year of seven thousand two hundred per acre, or two hundred and fifty billions for the state – a number which, placed one to each square inch of surface, would cover an area of forty thousand acres.

Estimates of the average number of insects per square yard in this state give us, at farthest, ten thousand per acre for our whole area. On this basis, if the operations of the birds were to be suspended, the rate of increase of these insect hosts would be accelerated about seventy per cent., and their numbers, instead of remaining year by year at the present average figure, would be increased over two-thirds each year. Any one familiar with geometrical ratios will understand the inevitable result. In the second year we should find insects nearly three times as numerous as now, and, in about twelve years, if this increase were not otherwise checked, we should have the entire state carpeted with insects, one to the square inch over our whole territory. I have so arranged this computation as to exclude the insoluble question of the relative value of birds and predaceous or parasitic insects, unless we suppose that birds eat an undue *proportion* of beneficial species.

. . . the average damage done by insects in Illinois amounts to twenty million dollars a year. These are large figures, certainly; but when we find that this means only about fifty-six cents an acre, we begin to see their probability. At any rate, few intelligent farmers or gardeners would refuse an offer to insure complete protection, year after year, against insects of all sorts for *twenty-five* cents an acre per annum; and we will, therefore, place the damage at one-half of the above amount-ten million dollars per annum.

Supposing that, as a consequence of this investigation, we are able to take measures which shall result in the increase, by so much as one per cent., of the efficiency of birds as an insect police, the effect would be a diminution of the above injury to the amount of sixty-six thousand dollars per annum, equivalent to the addition of over one and one-half million dollars to the permanent value of our property; or if, as is in fact a most moderate estimate, we should succeed in increasing the efficiency of birds five per cent., we should thereby add eight and one-fourth million dollars to the permanent wealth of the state, provided, as before, that birds do not eat unduly of beneficial species.

Figure 5.2: Excerpt of Illinois naturalist Stephen Forbes's (1880) article that likely makes the first "ecosystem services" calculation concerning the monetary value of birds' insect control. Essentially, he stated that birds eat lots of insects that incur monetary damage, and promoting more birds reduces that damage in a cost-effective manner.

Valuation concepts have recently turned to the idea of "ecosystem services."[5] As shown in Figure 5.2, within a naturalist's massive 1880 article on the things birds eat, Stephen Forbes makes what might be the first ecosystem service calculation. He assumed that there were three birds per acre in Illinois during the six months of the growing season, that each bird eats, at minimum, 20 insects per day, or 7,200 insects per acre per year, while citing birds that eat 10 times more than that number. He then argues that estimates for insect densities are 10,000 per acre and that the loss of the birds would increase insect densities by 70%, which, he said, ignores the fact that the birds don't distinguish insects that humans consider beneficial. He then asks what benefit would result if we could increase the efficiency of birds by 1% (by increasing their population), or their marginal benefit at that density, and he comes up with the number of $66,000 (in 1880 dollars) per year for the state of Illinois.

Updating these numbers a bit, but likely not their accuracy, we find that insecticide costs in Georgia in 2003 ranged from $4.50 to $60.00 per acre per year, excluding application costs.[6] At the same time, Georgia public health costs related to insects[7] — everything from doctors' fees for head lice to the cost of flyswatters — are set at over $209 million. Total dollar losses, including both control and damage, for the state of Georgia were estimated at more than $680 million, with mosquitos taking the number one spot at a cost of $122 million. Georgia's nearly 58,000 square miles, or 37 million acres, implies an average damage of $18 per acre (compared with Forbes's 56 cents damage per acre in 1880 dollars for Illinois).[8] Following Forbes's logic and 70% number, the benefit of birds in Georgia might be about $10 per acre in terms of insect control, both preventing damage and reducing treatment costs. This calculation means that regulations or programs that increased Georgia's bird populations by 50% could be valued as providing a benefit of, perhaps, $5 per acre, or $185 million to the state on an annual basis. Not a shabby number.

If similar numbers for damage held true for North Carolina's 31 million acres, an increased bird population would provide a $155 million statewide benefit, and for Durham County's 186,000 acres, a nearly $1 million annual benefit. How much is $155 million worth these days? In a sobering comparison, total retail sales in North Carolina during the month of February 2007 were $9 billion, with over $310 million collected on that amount in taxes, according to the North Carolina Department of Revenue. Increasing statewide bird populations by 50% equals collecting half-a-month's sales tax.

Trees make satisfying neighborhoods.

Correlated feature	Retail land use	Neighborhood satisfaction	Trees and shrubs	Income	Education level
Neighborhood satisfaction	**-0.28**				
Trees and shrubs	**-0.28**	**0.31**			
Income	**-0.33**	0.18	0.10		
Education level	-0.13	**0.22**	0.14	**0.45**	
Length of stay	-0.09	0.01	**0.23**	**0.37**	**0.22**

Figure 5.3: Photos display residential areas in College Station, Texas, with varying levels of vegetation (courtesy of Chris Ellis). The correlation table reveals connections between each pair of features (after Ellis et al. 2006). Bold numbers show statistically significant correlations. For example, comparing the bottom and top photos shows the **−0.28** correlation indicating increased trees and shrubs with decreased retail land use.

People generally value vegetation. A study in College Station, Texas, surveyed people to identify satisfying environmental features where they lived (Figure 5.3). Eight hundred surveys were mailed out, and 311 were returned. A subset of these surveys included 122 respondents living within 1,500 feet of retail land-use areas. Using this information, researchers teased out various environmental and socioeconomic aspects correlated with the satisfaction that people have with their neighborhood.[9] Satisfaction is simple: Trees and shrubs enhance people's satisfaction when they live near areas with retail land use.[10]

What's satisfaction worth? How much do people value cleaner air or cleaner water or the loss of species? Most approaches evaluate values of nonmarket goods through surveys. In some situations, surveys ask people of their willingness-to-pay (WTP), perhaps through higher taxes or entrance fees, for something like a nature reserve or bicycle paths or other environmental benefit. In other cases, people already have something, perhaps a certain level of environmental benefits or other "quality-of-life" feature, and studies evaluate people's willingness-to-accept (WTA) a shopping center or job in lieu of environmental degradation or poorer living conditions.[11] These surveyed people may or may not actually have to pay for the goods or accept poorer living standards at their stated values.

People are not terribly rational. One interesting study split people into two groups and asked one group how much money they would be willing to pay to subsidize emissions abatement from an animal rendering plant.[12] The other group was asked how much money they would be willing to accept to tolerate the noxious odors. In this study the WTP value was roughly $105, but the WTA value was an astonishingly larger $735.

Why aren't these two numbers the same?[13] They should be if people were rational. People should place a value on an obnoxious odor independently of whether they're paying to avoid it or accepting money to tolerate it. One classic example: Give one subject a coffee cup, then ask to buy it back (a WTA value — the subject thinks,"It's pretty and it's worth *so* much!"). Offer to sell another subject the identical coffee cup (a WTP value — "I have a lot like it, thank you very much."). This example reveals much about myself: I have a garage full of stuff I wouldn't pay a dime for in a store (it's WTP value is near zero), but I just can't seem to throw it away (it has a nonzero WTA value).[14] Overall, WTA-to-WTP ratios are highest for nonmarket goods, intermediate for ordinary private goods (stuff you'd sell at a yard sale), and lowest for money.[15] Certainly these results have implications for conservation easements.[16] How much does a property owner demand as compensation for development rights versus how much would said property owner pay for development rights?

People like neat trees, not messy forests.

Liked characteristic	Number of comments
Trees (so many, so big, different kinds)	92
Built features (shelter, playground, etc)	84
Neatness (trimmed, manicured, kept up)	84
Pretty (scenic, beautiful)	76
Park area	69
Water	69
Wildlife area (fish, birds, squirrels)	42
Looks safe	41
Natural beauty (not man-made, woods)	41
Good place to live	39
Road	37
Walking area (flat, could walk there)	33

Disliked characteristic	Number of comments
Disorderly (cluttered, messy, dirty, not kept up)	56
Weeds	55
Gloomy (too dark, too bushy)	43
Looks dangerous	41
Trees (too many, they look dead)	38

Detroit, Michigan

Figure 5.4: Liked and disliked features about open spaces in a black neighborhood in Detroit, Michigan (photos courtesy of Rachel Kaplan, after Talbot and Kaplan 1984). Humans favor the top two areas over the more natural bottom two areas. Overall, people found high value in open spaces.

What characteristics of urban nature do people value? The results presented in Figure 5.4 summarize interviews of 97 people living in three low- to middle-income neighborhoods in Detroit, of whom 97% were black (presumably two or three nonblacks).[17] People's likes and dislikes sometimes conflict with one another when it comes to features of open spaces. For example, both "neatness" and "natural beauty" appear in the list of liked characteristics. One comment by the scientists undertaking the study revealed the importance of underlying experiences.[18] People who always lived in the city — never lived outside the city — disliked woodsy creek images and disorderly places, whereas elderly folks, presumably with more rural experiences, had the exact opposite feelings. Conversations with Durham's urban foresters revealed that they receive demanding calls from citizens that natural areas be mowed and cut back upon the mere sighting of a snake.[19] Efforts to increase and manage urban vegetation and trees must account for people's disparate views.

Apparently, city people aren't aware that natural areas include, by definition, general disorder: fallen branches, tangled shrubs, and, yes, snakes, insects, and rodents. Overall, ecologists and environmentalists have an educational challenge ahead of them to teach citizens more about natural beauty and help them align urbanites' sense of beauty with the beauty that nature yields. Perhaps the motto for urban open space should be "as natural as can be under the circumstances."

As I mentioned earlier, "Adopt-a-Stream" approaches modified as "Adopt-an-Ecosystem" programs could help raise the nonmarket value of open space because the personal connections have worth in people's hearts. Even adopt-a-tree programs scaled down to single, neighborhood trees would help. I value the many trees I've planted through the years; many are still growing, and I still remember several trees that died through fault or no fault of my own. The value of these trees reflects that of my family's chickens, and the search for ecosystem services ought not dismiss nonmarket goods. Other citizens may initially treat trees like a coffee cup ("No thanks, I don't need trees because I have this pretty lawn") but once lawn-owners plant trees and begin to value them like pet chickens, perhaps their nonmarket value will increase.

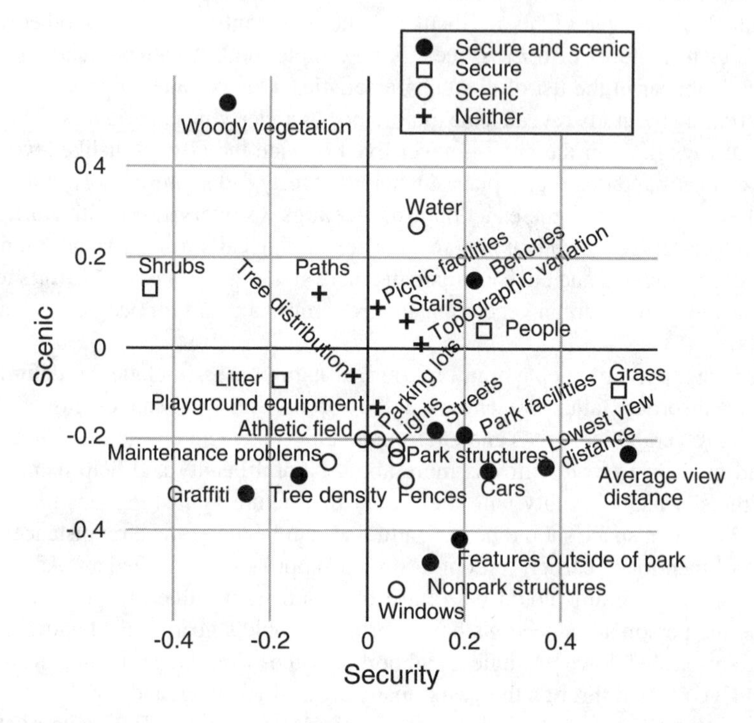

Figure 5.5: Geography, recreation, and psychology students rated photos of parks in Chicago, Illinois, and Atlanta, Georgia, in terms of security and scenic beauty (data from Schroeder and Anderson 1984). The above plot shows joint ratings for many different park features, and the symbols indicate the statistical significance of the correlations between the feature and the perceived aspect. For example, woody vegetation significantly correlated with both scenic beauty and *in*security, whereas seeing windows from the park correlated significantly with a lack of scenic beauty.

Even with high values for nonmarket goods, I find it difficult convincing even myself that a city council needs to spend money on those nonmarket feelings. Is a million dollars spent on urban vegetation worth a million dollars less spent on schools and public safety?

Another study also examined park features that people associate with scenic beauty, but extended it to uncover features providing a sense of security. Researchers took photos of 10 parks in Chicago, Illinois, and seven parks in Atlanta, Georgia, spanning the range from small to large, forested and lakefront, undeveloped and sports fields.[20] Some parks had unobstructed views of buildings, streets, and cars, while others were just heavily wooded. All photos used for Figure 5.5 were taken during summer daylight hours.

Three sets of a couple dozen students majoring in geography, recreation, and psychology, from Chicago, Atlanta, and East Lansing, Michigan, then rated these photos in terms of security and scenic beauty. Student ratings had a degree of repeatability: Some common photos were shown to various student groups, and the ratings of security and scenic beauty had significant correlations. Scientists performing the study analyzed the photos for lots of different features such as the fraction of the photo covered by grass or water, the presence of windows, visibility of structures outside the park, and many, many, more.

These results, at least through the perception of these students, mean that the most scenically beautiful park consists of woody vegetation around a pond or lake, along with benches, presumably on some paths. The safest park has grass with benches and long views with visible structures inside and outside the park, with plenty of people around. Ugly, unsafe parks have graffiti, high tree density, maintenance problems, and litter.

On the one hand, understanding how people value these various features provides valuable insight into park management: Parks must have benches! On the other hand, what do cities do with shrubs and woody vegetation? These features have great importance for wildlife, yet we see a direct and unfortunate conflict between natural vegetation and perceived security.

Underbrush was bad as far back as 1285.

And further, it is commanded that highways leading from one market town to another shall be enlarged, whereas woods, hedges, or dykes be, so that there be neither dyke, underwood, nor bush whereby a man may lurk to do hurt, near to the way, within two hundred foot of the one side and two hundred foot on the other side ; so that this statute shall not extend unto oaks, nor unto great trees, so as it shall be clear underneath. And if by default of the lord that will not abate the dyke, underwood, or bushes, in the manner aforesaid, any robberies be done therein, the lord shall be answerable for the felony ; and if murder be done the lord shall make a fine at the king's pleasure. And if the lord be not able to fell the underwoods, the country shall aid him therein. And the king willeth that in his demesne lands and woods, within his forest and without, the ways shall be enlarged as before is said. And if perchance a park be near to the highway, it is requisite that the lord shall minish his park so that there be a border of two hundred foot near the highway, as before is said, or that he make such a wall, dyke, or hedge that offenders may not pass, nor return to do evil.

— King Edward I, 1285

Figure 5.6: A section of the 1285 Statute of Winchester concerns the dangers of vegetation as a hiding place for evildoers (Adams and Stephens 1901, p. 78). "Demesne" means, essentially, domain, and "minish" is the same as diminish or reduce. The photograph shows a scene from Duke University's East Campus that lives up to King Edward I's ideal, with underbrush removed and "great trees" limbed for long sightlines.

For some reason, people think that vegetation protects criminals, harboring unknown dangers for the innocent, and this sense of danger goes way back. The 1285 Statute of Winchester[21] reveals that the idea that vegetation protects the criminal has been around for a very long time (see Figure 5.6). The list of disliked features in Figures 5.4 and 5.5 included this fear, along with cluttered and not kept up; the converse, "looks safe," along with trimmed and kept up, showed up on the list of liked features. Whether or not this statute from 1285 has any bearing, perhaps filtered through the style of English campuses designed centuries ago, Durham's East Campus of Duke University sets back its great oaks far from streets and walking paths, trimmed perhaps to reveal any offenders therein.

Is this fear of vegetation justified?[22] Eight hundred years ago criminals could run up to slow-moving carriages, but today's criminal-crunching cars weaken those fears. Certainly, bad (and good) people can hide behind many things, trees as well as parked cars, alleyways, brick walls, fences, shrubs, poles, thick grass, and dark hallways.[23] Would urban citizens' quality of life be better with empty, tree-free parking lots and highly urbanized areas, with roads cleared 200 feet on either side? We've already examined health issues relating to urban heating caused by replacing trees with parking lots. Bringing nature back into the city might alleviate those problems, and the remainder of this chapter considers whether having it directly promotes or hinders our safety and health. Whether or not King Edward's fear is justified today, it exists, and promoters of city vegetation must deal with the trade-off between the messiness of truly natural ecosystems and people's desires for trimmed trees without brush or sticks (perhaps motivated by perceived dangers).

Only a relative handful of studies examine these essential sociological questions. Descriptive examples exist in the social sciences literature regarding the fear of crime, or danger in general, but many lack quantitative analyses.[24] These examples include studies that rate the safety of parking lots and show that perceived danger increases with the amount of vegetation. Perhaps that perception explains a situation like the tree-free case of the Northgate parking lot (see Figure 2.12). Perhaps there is some basis to the fear. Interviews with informants concerning auto burglaries in Washington, D.C., for example, confirmed that criminals use available trees and vegetation to their advantage in a criminal act. Importantly, however, the commitment of the criminal act didn't depend on the presence or absence of trees and vegetation; rather, criminals just modified their behavior in the absence of vegetation.[25]

About 10 out of 10 people prefer malls.

Figure 5.7: Which spot is valued more? West Point Park on the Eno River holds the Eno Festival in July. Note the dozen cars at some random time. Compare that to the relatively new "Streets at Southpoint" mall with hundreds of vehicles (images provided by Durham County GIS, courtesy of Duane Therriault). (Note the lower-albedo reflective white-ish roof, surrounded by darker asphalt.)

Despite what people might state in a survey, in the end they reveal their preferences through their purchasing and recreational choices. The two images from Durham shown in Figure 5.7 demonstrate modern preferences by the number of parked cars. One image shows West Point Park on the Eno River, a park with natural and historical elements, and the other shows the relatively new Streets at Southpoint Mall (see Figure 2.1). Preserving and creating urban open space has two facets. On the one hand, open space preservation means having undisturbed natural spaces to enjoy, but, on the other hand, people appreciating those spaces must accept unfulfilled shopping desires.[26]

The need for ecosystem valuation really began with the 1989 *Exxon Valdez* oil tanker grounding in Alaska, combined with a 1989 court decision that allowed compensation for the loss of passive use values.[27] Despite the difficulties ascertaining passive use values, many such examples exist. In some situations, researchers set up "experimental markets": They describe and explain a nonmarket good, then have participants exchange the good in a laboratory setting. People's valuation of environmental features (in 1988 U.S. dollars) include $10.96 for farmland viewsheds, 36 cents per additional tree in a public park, $435 to increase air quality from "fair" to "good," and $273 to increase water quality to "boatable, swimmable, fishable."[28]

Other valuation approaches seek dollar estimates for what nature provides humans in terms of market goods and services. On a worldwide basis, annual valuation estimates for 17 ecosystem services include $17 trillion for nutrient cycling, $3 trillion for cultural value, $2.3 trillion for waste treatment, $1.7 trillion for water supply, and $1.4 trillion for food production. In total, the economic value of the world's ecosystems sits around $33 trillion annual productivity, though with a range of $16 to 54 trillion.[29] In comparison, sources estimate the world's gross product at around $18 trillion.[30] These ecosystem services, whether valued correctly or not, represent the interest gained on the world's natural capital. Sustainability concepts reflect the best of conservative values — preserving natural capital for future generations and not overspending it today.[31] Putting more and more stress on poorly understood ecosystems, to the point of major change, risks abruptly losing the services of this natural capital if ecological issues aren't addressed.[32]

Pushing the preservation of natural spaces primarily based on their economic value needs careful consideration, I think, simply because other uses will likely generate greater, more tangible, and shorter-term economic payoffs.[33] Living by the economic sword might mean further environmental setbacks by the economic sword.[34]

Varying tree cover in Chicago public housing.

Figure 5.8: Views of the Ida B. Wells homes in Chicago, Illinois (images by W.C. Sullivan, courtesy of F. Kuo; after Kuo and Sullivan 2001a). Top image shows a relatively tree-free view; bottom image shows a more "natural" view.

Over the next few pages I discuss studies performed at two public housing complexes in Chicago, Illinois. One of these, the Ida B. Wells development (see Figure 5.8), was built in 1941 *specifically* for African Americans.[35] The United States has had a long history of "poorhouses," resulting from a local responsibility of caring for the poor and elderly. During the late 1990s, the time of the study at this complex, the development housed nearly 6,000 people in 124 one- to four-story buildings. Researchers included 98 of the 124 buildings, leaving out buildings that, for one reason or another, were special cases.

According to housing development rules, staff assigned new residents to randomly chosen units. Random assignment means that people couldn't pick and choose buildings with more or less vegetation, not to mention more or less crime. This placement meant a simplification for a statistical study because individual outlook and attitude, level of vegetation, and incidence of crime started out independently. As the two photos depict, tremendous variation existed in the vegetation surrounding these apartment buildings. Also keep in mind that social spaces surrounding these buildings developed haphazardly; enhancing social interactions wasn't part of the environmental design of the buildings.

During the study, residents comprised 65% female, 97% African American, 93% unemployed, and 44% under the age of 14.[36] Serious social and economic strains make me think that vegetation was furthest from their many concerns. Still, using two years of crime reports from the Ida B. Wells homes, researchers correlated the type of crime with features of the location where the crime took place, including vegetation.

The Robert Taylor Homes was another public housing development in Chicago — the last remaining building was demolished in 2007 — and served as the site of two more studies involving people and vegetation, one exploring how vegetation calms people (Figure 5.9) and the other involving child development (Figure 5.10).

Again after special case exclusions, researchers examined 18 of the development's 28 16-story buildings, surveying people in floors 2 through 4 where residents might view vegetation outside their apartment windows. The study included a total of 145 women under 65 years of age, correlating their reported levels of violence with the greenness of the view from their apartment.

I show results from both housing development studies on the next four pages.

Reduced vegetation correlates with higher crime.

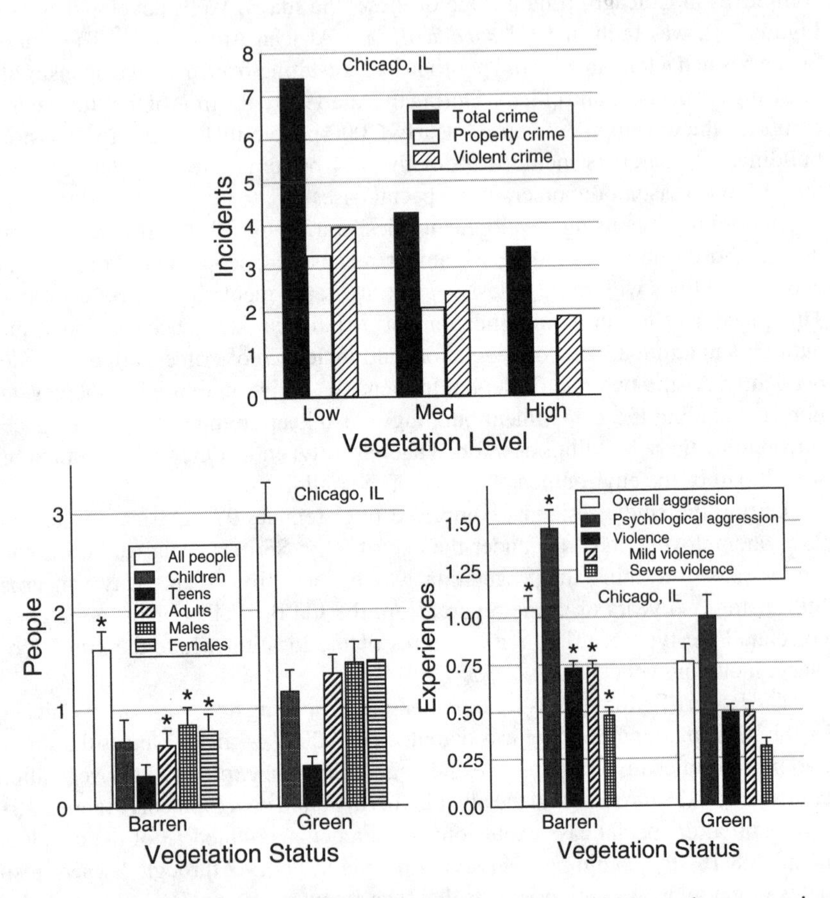

Figure 5.9: At top are results of a two-year study in the late 1990s on the connection between crime and vegetation at the Ida B. Wells homes (after Kuo and Sullivan 2001a). Crime is lower in areas of high vegetation. At bottom left, more people engage in social activities in vegetated areas at the Ida B. Wells homes in Chicago (after Sullivan et al. 2004). At bottom right, vegetation provides a calming effect, reducing the levels of violence against a partner and children (after Kuo and Sullivan 2001b).

Results of recent studies show that vegetation played a role in residents' lives at the Ida B. Wells public housing development. The values reported in Figure 5.9 are incidents per building over a two-year period, with 4 to 20 apartments per building, having an average occupancy rate per building of 7.8, ranging from 0.5 to 15. In particular, the reduction in crime between buildings with high and low levels of vegetation measures one incident for every two occupied apartments over a period of two years.[37] Vigilance likely drives this correlation: People use greener spaces more often, and this use increases vigilance and deters crime. In this vein, another study tested whether residents preferred vegetated areas.[38] Connecting measures of vegetation status to people's revealed preferences, lo and behold, showed that people — with the exception of teenagers — preferred green locations over barren ones. Social activities between women, like eating and talking in a group, nearly doubled in green spaces compared with barren ones. If vegetation promotes social activities, then, perhaps, the additional social activity provides protection against violence. A fine line apparently exists between people feeling afraid of vegetated areas (Figure 5.4) and feeling drawn to socialize in them.

Residents also experienced high levels of domestic violence: 61% reported aggression against their partner at least once in their lives, and 56% reported hitting a child, levels four times higher than the national average.[39] Results here indicate a correlation between a "green" environment and lower rates of domestic violence against a partner, while only "psychologically aggressive tactics" changed significantly between vegetation levels considering violence against children.

What mechanism connects domestic violence and vegetation? One hypothesis, called attention restoration theory,[40] proposes that all the stimuli in natural spaces — birds, bugs, flowers, the wind blowing leaves, and so on — involuntarily grab a person's attention, providing relief to the exhaustion of directed, voluntary attention. This diverted attention, so it proposes, provides a calming effect. One supporting point comes from pulling out the dependence of violence on attention abilities. Removing the correlation between more violent incidents and the interviewees' poorer performance on attention measures makes the link between vegetation and aggression disappear.

In dollar terms, the average loss stemming from all 23,440,720 U.S. crimes in 2005 was just $726, with personal crimes averaging $257 and property crimes $866.[41] Making no adjustments for inflation or family income, increased vegetation via reduced crime runs a few hundred dollars per occupied apartment per year, or about $2,900 per building. These crimes add up in a 124-building complex with almost 6,000 people. It seems that city leaders must balance additional police patrols with promoting more enjoyable spaces for watchful good people.[42]

Girls' self-discipline develops better with nature.

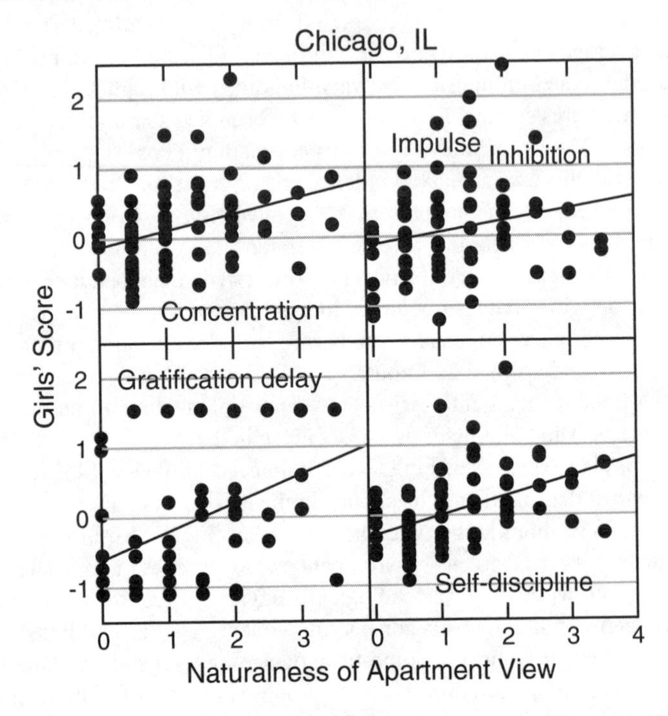

Figure 5.10: Girls experienced enhanced self-discipline development at the Robert Taylor Homes in Chicago, Illinois, with better views of vegetation from their apartment windows (after Taylor et al. 2002). Three separate measures — concentration, delay of gratification, and impulse inhibition — and their combined "self-discipline" measure, show statistically significant increasing trends with vegetation. Girls outperformed boys in concentration and impulse inhibition, whereas boys outperformed girls in delay of gratification. Boys' overall development showed no statistically significant trends with vegetation.

Does vegetation help children develop better? A study carried out at the Robert Taylor Homes (described on page 149) examined just that question.[43]

Researchers recruited four housing development residents, all of them African American women between the ages of 30 and 45, and trained them intensively to carry out one-on-one survey interviews with interviewees they had never previously met. Overall, the study included 169 interviews, with each interview involving both a mother (or primary caregiver) and a child. The children, all aged 7 to 12, included 91 boys and 78 girls. Furthermore, during the interviews, child-care providers entertained children who were not involved in the interview, removing this child-care distraction.

Interviewees rated nature, as viewed from their apartment, on a five-point scale.[44] Concentration involved four tasks — substitutions, sequences, reverse alphabet recitation, and image recognition — combined into a single score. Three tasks measured impulse inhibition: matching detailed and similar images, reciting color names printed in incongruent colors, and matching conceptually similar icons. Each task involved resisting an impulse to quickly provide answers that demand careful consideration. Finally, interviewers measured "delay of gratification" using small and large bags of a child's preferred candy: The interviewer told the child that he or she could have the large bag if the child could wait long enough (up to a maximum waiting time of 15 minutes), then told the child to close his or her eyes and take the candy out of the room. The child's delay of gratification score related to how long he or she could wait for the large bag of candy.

In summary, "green views" from a girl's window enhanced her development, at least as measured by the four scores studied here, but boys' development did not. Overall, girls exceeded boys at concentrating and inhibiting their impulses (see Figure 5.10), although boys exceeded girls in the delay of gratification study, with girls waiting an average of 358 seconds versus 454 seconds for boys. Thus, boys could delay their gratification longer, but the delay did not depend on vegetation. There was one explanation for the difference between boys and girls: Some evidence shows that boys play further away from home than do girls, and are less affected by near-home nature.

The study points out one cautionary note, among several: Mothers with a positive outlook might rate their views greener (more positively), and because of their positive outlook, their children have a more positive development. Thus, the mother's outlook correlates the rating of nature and the development of girls. However, boys should also be affected by this correlation.

Nature promotes emotional and physical health.

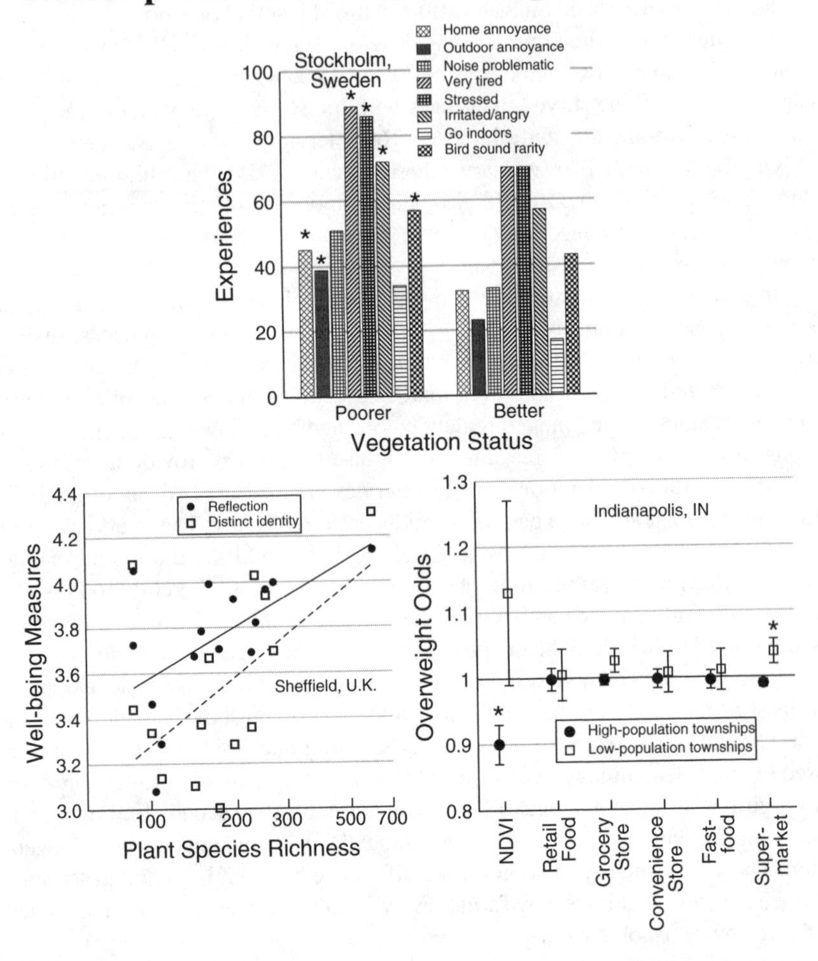

Figure 5.11: The top chart shows the frequency of various feelings by residents of Stockholm, Sweden, with noise on both sides of their apartments (after Gidlöf-Gunnarsson and Öhrström 2007). Better access to vegetation relieves a variety of negative feelings. At bottom left, two measures of well-being increase with plant species richness (after Fuller et al. 2007). At bottom right, living in more vegetated neighborhoods (measured through NDVI) reduces the likelihood of childhood obesity in high-density areas (after Liu et al. 2007).

At top of Figure 5.11, a study from Stockholm, Sweden, examined how access to green areas affected residents with and without quiet space available at their apartments, relating 500 carefully selected questionaire responses to environmental conditions.[45] All residents faced a road on one side, and both sides of their apartments had noise levels around 60–68 decibels. Along with questions concerning how noise affects them, they were asked about their access to green areas. In this plot, a star above a column indicates that better access to green areas has a statistically significant effect on the experience, although these effects weren't necessarily large ones. For example, concerning annoyance at home and outdoors, access to green areas explained only about 3% of the variation. However, many features were affected positively by access to green spaces.

More than just "green," people feel better with something that correlates with species richness. Here I show results, at bottom left, for two "feel-good" measures against plant species diversity in Sheffield, UK.[46] Researchers define *reflection* as the "ability to think and gain perspective" and *distinct identity* as the "degree of feeling unique or different through association with a particular place." Responses ranked these feelings on a five-point scale, being elicited during interviews from 312 users of 15 different greenspaces ranging from 1 to 24 hectares each. Also important were the number of distinct habitats within a greenspace, as well as the number of bird species present.

In addition to emotional health, a childhood obesity study from Marion County, Indiana, examined the medical records of more than 7,000 children between 3 and 18 years of age, in the different townships. Defining obesity as above the 95th percentile for their age, children living in high-density townships had lower odds of obesity when living with higher vegetation, as summarized at bottom right. In low-density townships, children living further from a brand-name supermarket had higher obesity odds. A clear mechanism exists for the former association — children with vegetation play outside more often — but the latter mechanism seems elusive. Another study in Indianapolis, Indiana, included nearly 4,000 children aged 3–16, mostly Medicaid enrollees from lower-income families. Over a two-year period, researchers observed a highly significant connection between a child's body mass index (BMI) and the normalized difference vegetation index (NDVI) where that child lived. In that study the odds-ratio was 0.87, with a 95% confidence interval of (0.79–0.97), translating to decreased body masses with increased vegetation of 1.6 kg and 5.1 kg for girls aged 4 and 16, respectively.[47] Researchers presume that greater activity levels provide the mechanism.

Trees promote bird and plant species richness.

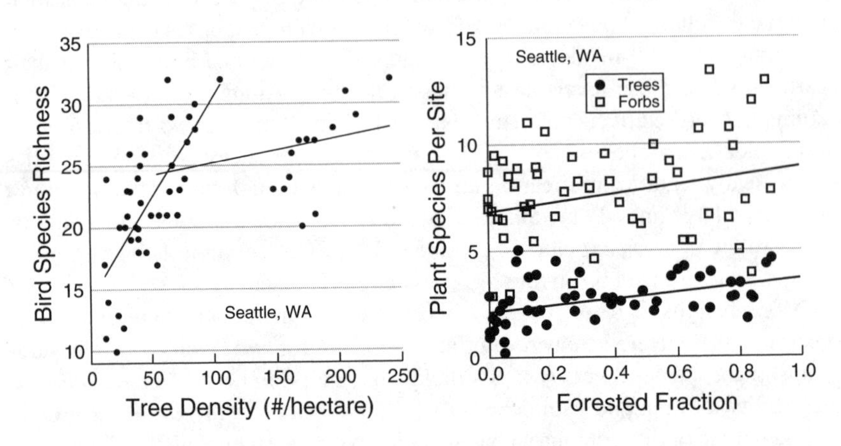

Figure 5.12: At left, censusing birds in developed areas of Seattle, Washington, demonstrate increasing species richness with tree density (after Donnelly and Marzluff 2006). In that area, if you want birds, you need a minimum of 10 trees per hectare. At right, another look at Seattle connected plant species surveys with measures of forested areas in 1 square kilometer parcels, finding that areas with a greater fraction of forest cover have significantly more tree and forb species (after Hansen et al. 2005). Shrub species, however, didn't increase significantly with forested fraction.

Nature also values nature. The previous set of plots showed that people simply enjoy seeing a diversity of birds and trees, and the graphs shown here in Figure 5.12 demonstrate, at least in urban Seattle, Washington, how bird and plant species richness depends on tree and forest abundance.[48] At left, the two different spans of tree density indicate that, at low tree density, every 6 trees per hectare (about 2.5 acres) adds another bird species. At higher densities, above 50 to 100 trees per hectare, every 50-some trees adds a new bird species.[49] Lots of trees also beget lots of plant species, here again shown for Seattle. Tree species diversity increases with forest cover, as well as the forbs, which essentially constitute all herbaceous flowering plants, excluding grasses.

If you value birds and want them in the city, then residential lots need an absolute minimum of 10 trees per hectare, or about 5 trees per acre, with about one-quarter of these trees being evergreens.[50] In other words, if you have a one-quarter acre lot (about 100 feet by 100 feet), then you should have at least one evergreen and one deciduous tree. Better yet, have five of each.

Of course, birds don't care about the boundaries of city lots, so having a couple of trees on a one-quarter acre lot in the midst of a treeless subdivision of manicured lawns won't do much for the birds. Promoting high urban bird populations becomes a community endeavor, a public good, something that should done by homeowners for the greater good of nature, as well as a public health benefit via insect control (see Figure 5.2).

Keep in mind that not every square meter should be turned into tree canopy: Some bird species require early successional grassland environments (also vastly different from a manicured, pesticide-covered lawn). As evidence, increasing urbanization in Boulder, Colorado, reduced the richness of bird species preferring grasslands,[51] and many bird species went from present to absent when urban land-cover fractions exceeded a value around 50%. In such a situation, cities — at least cities that want birds — should be kept below 52% urban landcover, with some forest patches of at least 42 hectares in extent.

As far as patches go, wildlife needs multiple reserves that serve as sources and sinks for colonization and extinctions.[52] Whether or not parks and treed neighborhoods, along with stream corridors, serve as multiple reserves, no one really knows. Until scientists better understand such practical urban ecology concepts, just make the guess that more reserves are better than fewer reserves.

Chapter 6

Human Health and Urban Inequities

People create cities having a peculiar range of unnatural environmental conditions, shown in previous chapters, and here I cover health-related heat and pollution in cities. I then provide examples of the stark socioeconomic inequities of these conditions, mirroring inequities in the provisioning of urban natural areas.

Health problems arise from both heat and air pollution, with one dramatic example being an extreme heat wave that increased mortality rates in the elderly population. More subtle problems arise with air pollution, which increases heart attacks in the summer and pneumonia in the winter. Asthma represents another chronic problem, with high levels of either ozone or sulfur dioxide increasing the odds of suffering childhood asthma by 50% or more. Still, plots of asthma incidence and pollutant levels don't have a clear signal; that is, the connections between the two issues are hard to uncover.

Given these urban environmental problems, one must still balance them against fundamentally different age-dependent mortality patterns for different genders and socioeconomic groups when seeking solutions and allocating scarce resources. Air conditioning provides a technological solution that clearly reduces heat-related mortality, but even that simple measure shows an inequitable socioeconomic distribution. Also bear in mind the results from previous chapters showing heat and energy use in cities: Air conditioning use sharply increases our peak demand on hot days, potentially exacerbating the problem with emissions from power plants.

This chapter also explores the variation in a number of socioeconomic variables, along with distributions of urban environmental features, such as urban nature and aspects of health. Several specific studies demonstrate vegetation inequities based on income in Durham, North Carolina, Milwaukee, Wisconsin, and Baltimore, Maryland. This inequitable distribution of vegetation extends to features correlated with income, such as rentership and education. Mechanisms underpinning these inequities have many origins, sometimes going back to landscaping decisions made decades earlier by previous homeowners.

Inequities in urban nature go beyond private "holdings" of vegetation to low levels of public park access for poor people in Los Angeles. Perhaps more surprisingly, cases show that wealthier people live in areas with higher plant species richness, meaning an inequitable distribution of biodiversity. As shown in the previous chapter, clear benefits result from having views of and access to nature, meaning that all of these environmental inequities, taken together, lead to other inequities, including physical and emotional health. It remains a question for urban areas to address how important, or strong, these benefits are within a broader context of urban citizens' lives.

Heat waves lead to deaths a few days later.

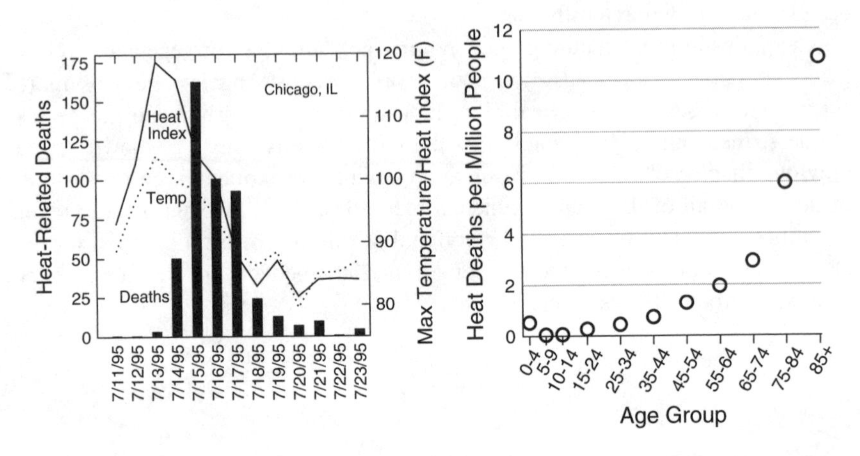

Figure 6.1: A heat wave swept Chicago, Illinois, during July 11–23, 1995, leading to at least 700 heat-related deaths (left plot). I plot both the temperature and the heat index, a measure of "effective temperature" that accounts for reduced perspiration effectiveness because of high humidity. The peak of heat-related deaths occurs several days after the heat wave (after McGeehin and Mirabelli 2001). A study by the Centers for Disease Control shows, at right, that our elderly population risks the greatest heat-related mortality (after CDC 1995). Underlying these mortality rates were 5,379 deaths attributed to heat between 1979 and 1992. The CDC data indicate, for example, around 10 deaths per million 80-year-old people, against a background mortality of roughly 70,000 deaths per million 80-year-old people (see Figure 6.7).

Most heat-related deaths occur from heatstroke, defined by a body temperature exceeding 105F (40.6C), compared to normal body temperature of 98.6F (37C). Variability causes the problems, rather than heat itself (see Figure 6.5), with duration, overnight heat, and non–air-conditioned automobiles and dwellings appearing critical (see Figure 6.6).[1] Heat affects urban dwellers most — because of the urban heat island — and in particular, the young, elderly, poor, and people on certain medications. Even factors like living on the top floor of an apartment without air conditioning affects one's risk. Clearly important strategies that reduce this heat-related risk include drinking liquids and having air conditioning, connecting the availability of coping strategies to socioeconomic status.

A nasty heat wave that hit Chicago, Illnois, in 1995 clearly demonstrates heat-related mortality and how that mortality takes place a few days after the heat wave's passing. During just few days, July 11-23, there were 85% more deaths, and hospitalizations increased by 11% compared to the same period the previous year. In total, more than 700 additional deaths took place because of heat. I have not seen any clear mechanism spelled out explaining the time lag between heat and mortality, but it may not be that big of a puzzle: Perhaps people think they can manage the heat, and do so for a while, but then collapse from stress and dehydration.[2]

The heat index combines temperature and humidity, generally increasing with both variables, and better represents the body's cooling ability through perspiration. The U.S. National Oceanic & Atmospheric Administration (NOAA) has a chart showing dangerous heat index values: Any temperature above 100F and 40% humidity is either dangerous or extremely so, and a humidity of 100% is dangerous for temperatures above 90F. A mathematically complicated and slightly curved line joins these two points, but a simpler straight line doesn't do too bad a job. The heat index's dangerous limit for humidity R (measured as percent) depends on temperature T (in Fahrenheit) as $R = 100 - 6(T - 90)$.[3] During the summers in Durham, I certainly can attest to this line being roughly correct.

The Centers for Disease Control (CDC) analyzed more than 5,000 deaths between 1979 and 1992 that were listed as heat-related on their death certificate (excluding listings identified by conditions aggravated by heat, but not explicitly stating heat). By their estimate, between 148 and 1,700 heat-related deaths occur each year, with Alabama, Arkansas, Arizona, Georgia, Kansas, Mississippi, Missouri, Oklahoma, and South Carolina listed as the states with the greatest mortality risk.[4] The CDC analysis demonstrates that increasing age presents the greatest risk factor associated with heat-related death, though relative to other children, children less than 4 years old also have a higher risk.

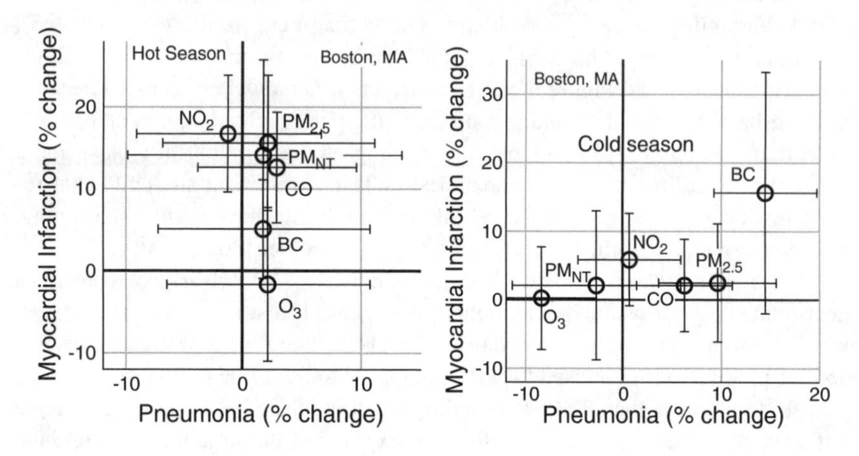

Figure 6.2: A large study from Boston, Massachusetts, connected air pollution levels to hospital admission rates of patients aged 65 years and older, an age group most susceptible to the effects of heat (see Figure 6.1). Dividing the data into "hot" (left) and "cold" (right) seasons, these plots present the percentage changes between hospital admissions between periods of high and low levels of the various pollutants (after Zanobetti and Schwartz 2006). For example, in the cold season, high levels of black carbon (BC) correspond to about 15% higher admissions for both myocardial infarction (heart attack) and pneumonia when compared to low levels of BC. Error bars show the 95% confidence intervals.

The two plots in Figure 6.2, one for hot and one for cold seasons, examine emergency department admissions from 1995 to 1999 for 15,578 heart attack (myocardial infarction) patients and 24,857 pneumonia patients aged 65 and older.[5] Researchers matched each of these emergency room admissions with the pollution level on admission day. Pulling out days with the lowest and highest pollution levels — specifically the bottom and top 10% — they tested whether there was a difference between admission rates in these tails of the pollution distributions. Further, they separately examined differences in excess admissions for "cold" (October to March) and "hot" (April to September) seasons — lots of data and lots of tests.

The short answer: Air pollution exacerbates pneumonia in the winter and heart attacks in the summer. For example, the plots show that black carbon in the hot season causes no difference in admissions for pneumonia, but comparing the lowest and highest pollution concentrations in the cold season shows that black carbon causes about a 15% increase in hospitalizations. Error bars show the limits that demarcate the 95% confidence interval, a standard level of significant effect. What does that mean? Suppose the world went back in time to the beginning of the study period, and, although the pollution levels played out the same, everybody got a new chance at getting struck with pneumonia or a heart attack. If this rewinding happened 20 times, statisticians expect that measurements from 19 of those rewindings would fall within the 95% confidence interval (0.95 = 19/20). Coming back to black carbon in the cool season, one is tempted to see it correlated with heart attacks, but that bottom limit sneaks just below zero. This temptation induces scientists to find larger databases with more cases, hoping to reduce the confidence interval, perhaps revealing such a correlation as significant.

Pneumonia in winter significantly correlates with black carbon (BC), carbon monoxide (CO), and $PM_{2.5}$. Summertime pneumonia hospitalizations show no connection to any of the pollutants. Regarding heart attacks, CO had the greatest effect in the summer, along with NO_2, $PM_{2.5}$, and nontraffic particulate matter, labeled PM_{NT}.

Although ozone appears to reduce pneumonia incidence in winter, reduced ozone levels correlate with higher levels of the other pollutants, which correlate with higher hospital admissions rates. Perhaps this correlation comes through here. Or perhaps ozone provided some protective benefit from pneumonia to this population. Whatever is going on here, the next study shows an example of harmful ozone.

High ozone and SO_2 levels predict high asthma hospitalizations.

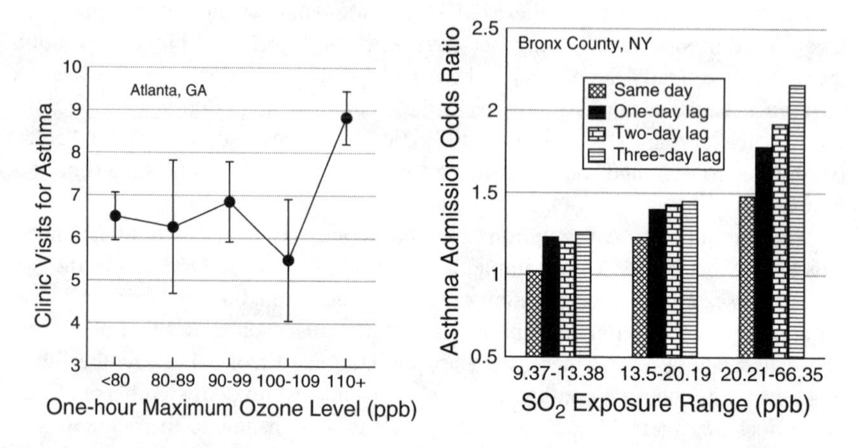

Figure 6.3: At left, daily asthma admissions for children, aged 1 to 16, in an Atlanta, Georgia, public hospital during the summer of 1990 showed a marked increase when one-hour ozone levels exceeded 110 ppb (after White et al. 1994). The right-hand plot summarizes asthma admissions for children aged 0–14 years old from Bronx County, New York, between June 1991 and December 1993. The odds of being admitted to the hospital for asthma differ between the four quartiles (dividing the distribution into four equal parts) of the SO_2 concentration distribution, with admissions rates for the lowest quartile defining admission odds of 1.0 (after Lin et al. 2004). Not only do higher pollutant levels increase admission odds, but odds depend most on the SO_2 concentration three days prior to admission.

Air pollution causes an array of health problems, including heart problems, cancer, and altered immune responses.[6] Exposure to ozone and allergens can cause asthma, an inflammation of the small (<1 mm) bronchioles in the lungs. The upper respiratory tract absorbs, for example, highly water-soluble sulfur dioxide (SO_2), and the subsequent irritation constricts the airway.[7] These exposures affect children in particular because of their less-developed lungs and immune systems.[8]

Figure 6.3 demonstrates how two types of air pollution affect the frequency of asthma attacks. One dataset (the left plot) from the Grady Memorial Hospital in Atlanta, Georgia, a hospital that served a low-income, predominantly black population, shows that the highest ozone days have 37% higher pediatric asthma hospitalizations. Unfortunately, the results come from a small sample of patients, just 609 hospital visits for children aged 1 to 16; this means that the results have relatively weak statistical strength. The other plot, on the right, shows that in Bronx County, New York, childhood asthma admissions and SO_2 data have a similar pattern when adjusted for race, age, and season.[9] In New York, 8.6% of children aged 0–17 suffer from asthma, substantially more than children in other low-income areas in the United States. Between the lowest and highest quartiles, asthma admissions odds increased by a factor of 1.48 to 2.16 based on average daily SO_2 levels, and 1.55 to 1.86 based on maximum daily SO_2 levels. Admission rates also depend on the lag between high pollution day and admission day.[10]

Other studies from Atlanta confirm these results. Ozone levels above 100 ppb increased the odds of asthma hospital admissions one day later by 23% over admission rates when ozone levels were below 50 ppb. Also observed were large increases in asthma admission odds for PM_{10}.[11] More recent studies examined massive numbers of emergency room visits, more than 4 million, from 1993 to 2000.[12] Of these visits, 11% involved respiratory problems, around 172 cases per day. Analysis of combined data for all respiratory diseases, chronic obstructive pulmonary disease, upper respiratory infection, asthma, and pneumonia showed that increasing levels of pollutants, for example, 24-hour PM_{10}, 8-hour O_3, and 1-hour NO_2, by one standard deviation increased the incidence of respiratory disease by 1.3%, 2.4%, and 1.6%, respectively. Statistical analysis including all pollutants simultaneously gave a tantalizing result that only ozone had any real importance.[13]

Perhaps a better measure of pollution levels were the health implications. During the 1996 Atlanta Olympics period, data from four different sources[14] regarding 1- to 16-year-old children showed reductions in acute asthma events ranging from 11 to 44%, representing a total of 56 fewer admissions during that period. As a control to compare with the asthma admissions, these same sources found changes in nonasthma events ranging from -3.1 to 1.3%.

Asthma incidence and pollution aren't tightly correlated through time.

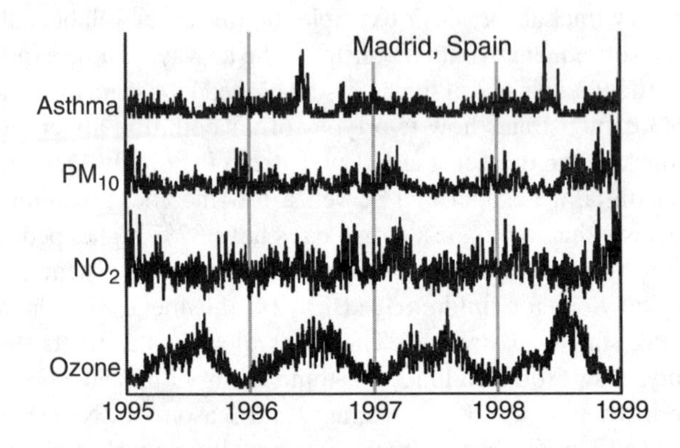

Figure 6.4: Temporal dynamics of pollutants and asthma in Madrid, Spain, from 1995 to 1998 (after Galan et al. 2003). These curves include two asthma epidemics that took place in May 1996 and May 1998, and show the subtle nature and relative weakness of the correlation between asthma and air pollution. In contrast, compare these plots of weakly related variables with tightly related variables, such as the strong 400,000-year correlation between CO_2 and temperature in Figure 3.6.

Although the data in Figure 6.3 give a clear mental picture of cause and effect, the connection between health issues and pollution isn't always so sharp. Figure 6.4 summarizes results from Madrid, Spain's emergency room admissions for asthma and various pollutant levels over a four-year time span.

When asking how respiratory diseases depend on pollutant levels, this plot demonstrates that the connection isn't perfect. For example, take a peek at the long-term connection between global CO_2 and temperature in Figure 3.6. That plot really needs no accompanying statistical analysis revealing whether or not a correlation exists. Weak correlations like the ones displayed here demonstrate the need for statistical analyses: Certainly the asthma curve doesn't simply reflect any of the other curves.

Despite the lack of a strong visual connection, statistical analysis indicates that PM_{10}, NO_2, and O_3 correlate positively and significantly with asthma events, confirming the results of Figure 6.3. A one-day lag in ozone and three-day lags in PM_{10} and NO_2 correlate most stongly.

Yet another study, this time in Taiwan in 2004, found significant correlations between emergency room cases of childhood asthma and NO_2, CO, and PM_{10}.[15] In that particular situation, ozone level was not a significant contribution for childhood asthma, and nothing reached significance for adult cases.

Other things besides pollution increase the risk of asthma attacks, including humidity and a rapid decrease from high barometric pressure. One estimate shows that these climatic variables provoke more than 20% of asthma attacks.[16] The asthma-inducing mechanism behind a rapid pressure decrease may be an increased pollen release corresponding with thunderstorms.[17] Incidentally, these studies also considered pollen as a factor for asthma (with ragweed pollen increasing in urban areas; see Figure 3.8), but those results didn't change the importance of the pollutants. Urbanites also suffer from asthma due to exposure to cockroaches (in the northeastern United States), rodents, dust mites (in the south and northwest), pets, and molds. Living with violence might also increase stresses that promote asthma.[18]

In summary, these results demonstrate that asthma has a complicated and messy dependence on air pollution, and ozone isn't the only pollutant that matters, despite my focus on it here.

Though cities differ, heat kills people in July and August.

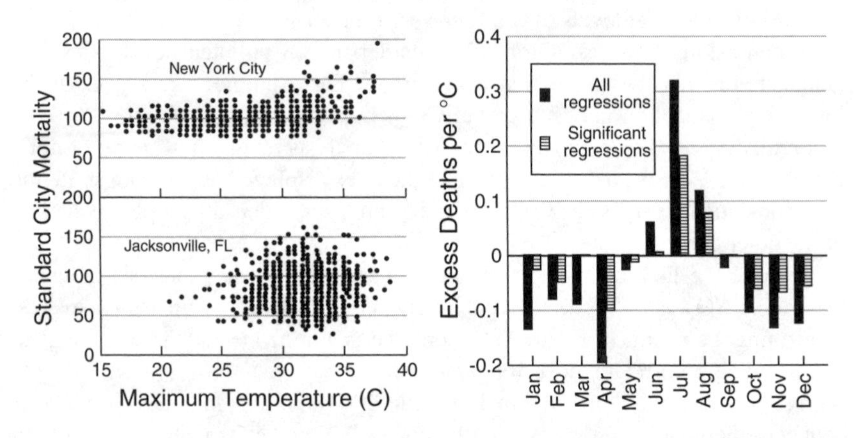

Figure 6.5: At left, plots from New York City, and Jacksonville, Florida, summarize data from the 1960s, 1970s, and 1980, showing how mortality varies with maximum temperature. Floridians seem to deal better with high heat, an idea borne out by a broader examination of regional mortality patterns (after Kalkstein and Davis 1989). The graph at right displays excess heat-related deaths per °C using data from 28 metropolitan areas over the years 1964–1998 (after Davis et al. 2004).

Plots from Jacksonville, Florida, and New York City, along with many other U.S. cities, combine data on deaths during the years 1964–1966, 1972–1978, and 1980 (all years lumped together) and consider them against temperature and other environmental variables, such as airspeed, dewpoint, and degree cooling and heating hours (Figure 6.5).[19] In order to compare death rates in all of these socioeconomically and geographically disparate cities, scientists calculate a standardized mortality to facilitate this cross-comparison (see Figure 6.7).

Plotting this standard mortality against temperatures for each day for a northern city and a southern city, in the left plot, reveals some evidence behind the idea of acclimating to heat. If a city doesn't experience extreme temperatures, then heat waves are quite problematic, but if a city is always hot, then what's a few degrees more? Citizens of hot cities apparently know how to deal with heat better than people in cities with more variable temperatures.

Indeed, two summer weather systems — dry tropical air masses and very warm and humid tropical air masses — in cities exceeding 1 million people cause significant mortality increases.[20] Total estimated deaths attributed to summer heat in 44 U.S. cities exceeded 1,800 per summer season, with the elderly being more susceptible. Furthermore, climate change models predict two to six times greater frequencies of these air masses above current levels, with expectations for overall mortality increasing by a factor of up to three.

Data regarding monthly mortality come from the National Center for Health Statistics, covering the years 1964–1966 and 1973–1998 for 28 metropolitan areas: 29 years worth of daily data on mortality and temperature.[21] Think about all the problems faced when comparing death rates in different cities in 1950, 1975, and 2000. Not only did the distribution of ages change differently in different places, so did the technologies associated with health care and automobile emissions regulations. Here is just one complicating example: Increased air conditioning from the 1980s to the 1990s greatly reduced mortality (see Figure 6.6). The scientists doing this work had to deal with these problems while untangling their question of how higher temperatures affects mortality. Let it suffice that there are "accepted" ways to standardize these mortality rates (see Figure 6.7).[22]

In the right-hand graph, after applying these standardizations, we see the expected increase in heat-related deaths, measured against a one degree Celcius temperature increase.[23] This plot's message: Summer heat is a bad thing, winter heat is a good thing.

Air conditioning reduces heat-related mortality.

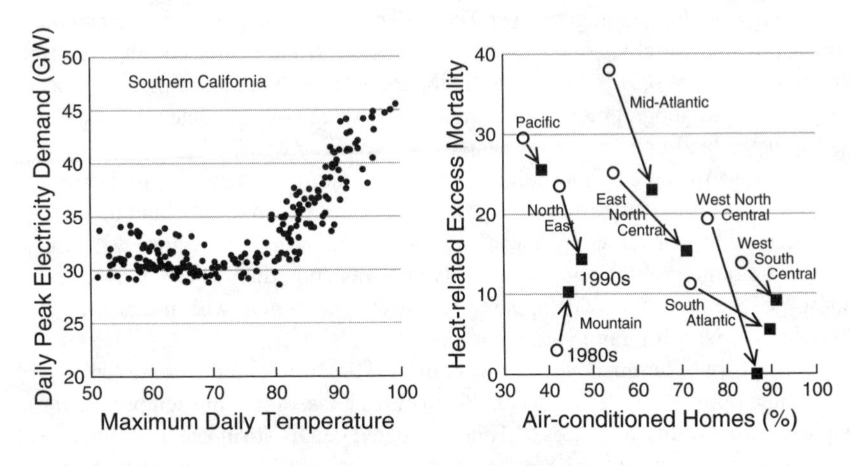

Figure 6.6: The left plot shows the amount of electricity supplied by Southern California Edison in 2004 versus the maximum daily temperature, with every day being a separate data point (data provided by Guido Franco, after Franco and Sanstad 2006). Electricity use increases when air conditioners turn on at high temperatures. At right, air conditioning reduced heat-related mortality (after Davis et al. 2003). Open circles indicate values from the 1980s, and the filled squares denote values from the 1990s. Most areas reduced heat-related mortality with increased air conditioning.

Let's put our energy use into the context of just one of the reasons for using it: Air conditioning relieves heat discomfort and improves health. We see the signal of AC use in California's peak electricity use, in the left-hand plot of Figure 6.6, increasing sharply as the temperature rises. As this California data shows, millions of air conditioners kick in on days when peak temperatures hit 80F. Here, the graph plots each day's peak electricity use against each day's peak temperature for every day from June to September 2004.[24] Electricity demand increases by about 3% for each degree Celsius above 27°C, or 80F. An increase in air conditioning demands more electrical power plants, more energy use, and all the associated local and global environmental problems.

Urban environments enhance the need for air conditioning: Greater urbanization increases the heat island effect and requires more electrically supplied air conditioning. Cooling the additional 0.5–3C temperature increase due to the urban heat island (UHI) effect uses an estimated 5 to 10% of urban energy. That cost reflects real money: UHI-associated cooling costs run about $2–4 billion per year over the United States.[25]

We cool our air not just because it provides comfort. The right-hand graph shows that our heat-related mortality decreased as more homes became air conditioned between the 1980s and 1990s.[26] This mortality reduction ignores the consequences of possible emissions from electricity production specifically for air conditioning. Certainly we use more air conditioning in the hottest times of the year, and, as we've seen in Figure 4.8, emissions from this increased use travel downwind and generate ozone, leading to greater health problems.

Should we reduce our energy use by reducing air conditioning, while accepting more asthma emergencies and heat-related deaths? Air conditioning cost estimates for the United States in 1999 were $36 billion per year, or about $100 per person per year.[27] Greater use of energy conservation strategies would reduce our need for air conditioning while simultaneously maintaining the health benefits, for example using thermal mass to minimize peak temperatures, as described in Chapter 2. Of course, these short-term estimated expenses ignore hard-to-calculate long-term costs of greenhouse gas emissions from fossil fuel use.

Finally, results shown here involve the standardized, age-adjusted mortality discussed in Figure 6.7, a standardization that seemingly magnified a mortality reduction trend compared with gross statistics. Whether or not that magnification is real, the air conditioning–mortality results shown here removed that trend before calculating heat-related excess mortality.

Many people die in winter (accounting for age, race, and gender).

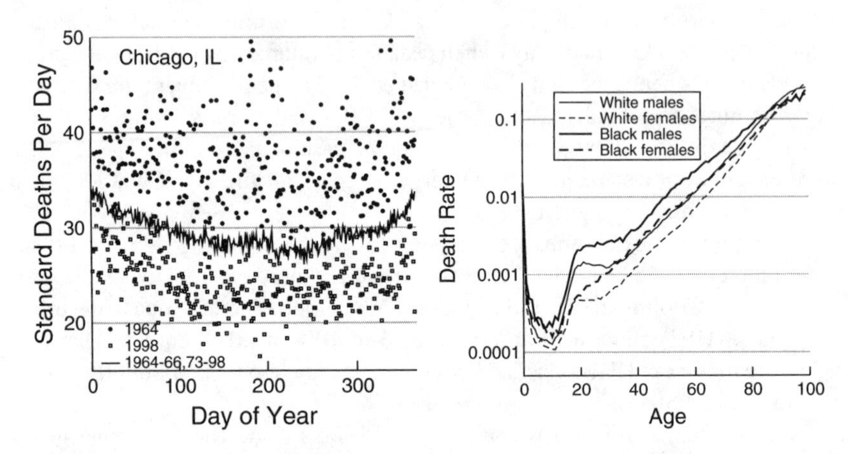

Figure 6.7: At left I show standard deaths per day versus day of year for Chicago, Illinois (data courtesy of Chip Knappenberger). Day 1 denotes the first of January, and Day 365 denotes December 31. These numbers relate to a "standardized" population of 1 million people. Data points indicate two different years, 1964 (filled circles) and 1998 (open squares), and the solid line gives the long-term average over 1964–1968 and 1973–1998. Winter mortality exceeds summer mortality, while standard mortality has greatly declined since the 1960s. At right, the graph displays raw mortality rates for white and black men and women in the year 2005, dividing the number of deaths at each age in that year by the July 2005 population of that age (data from the CDC and U.S. Census).

I want to place heat-related mortality into an appropriate context. Determining heat-related mortality in the previous plots relies, in some cases, on standardized mortality, and I show one such example of standardized mortality for Chicago, Illinois.[28] Differences in age distributions, health technology, and other variables across years and places means that mortality data must be "corrected" to make a meaningful comparison. These data show that winter mortality, scaled as daily deaths per million people, exceeds summer mortality. It's not clear why. Perhaps during summer heat, old people stay indoors rather than playing outside with chainsaws. Perhaps in winter darkness, people stay inside more, passing around infectious diseases. Seasonal behavioral choices like these might result in summer heat promoting safer indoor activities. If indeed heat changes behavior, conclusions regarding global warming and mortality rates in the coming decades become complicated by how we'll behave.

What mortality rate do we expect for a city of 1 million people? If a person lives for 80 years, that's a total of 29,220 days of life. If every day were like every other day, dividing 1 million by 29,220, we should expect roughly 34 deaths each day (and, happily, 34 births). This rough number doesn't work out too badly compared to the y-axis of Figure 6.7. Still, I'm a little uneasy. The average daily standardized mortality during the 1990s is just 26 deaths per day. Dividing 1 million people by 26 deaths per day gives a lifetime of about 38,500 days, or roughly 105 years of age. I know that number is too high.

Looking at actual mortality rates a bit deeper, the National Center for Health Statistics reports that the crude death rate (meaning the actual number of people who died in a year, not a *corrected* number) for 100,000 people in 2004 was 816.5, down from 963.8 in 1950. In contrast, standardized mortality rates differed greatly; in 2004 it was 800.8, but in 1950 it was 1,446.0. Surprise, surprise: We live longer now than in 1950, but standardized mortality makes the picture really rosy. Why? The underlying death rates for different ages are tremendously different. Indeed, the mortality rate plot on the right, which simply divides deaths by populations of specific categories of age, race, and gender, shows how drastically these values differ between young and old, male and female, black and white.[29] Another way to describe these differences is as follows: In 2004 the crude death rates for 100,000 people in age ranges 5–9, 45–49, and 75–79 were, respectively, 14.7, 352.3, and 4,193.2. Those are drastically different values, and any changes in the distribution of ages, whether because of a change in time or location, makes comparing death rates a tough problem.[30]

Also note the high mortality rates for the elderly and the rather small contribution from heat-related deaths (see Figure 6.1).

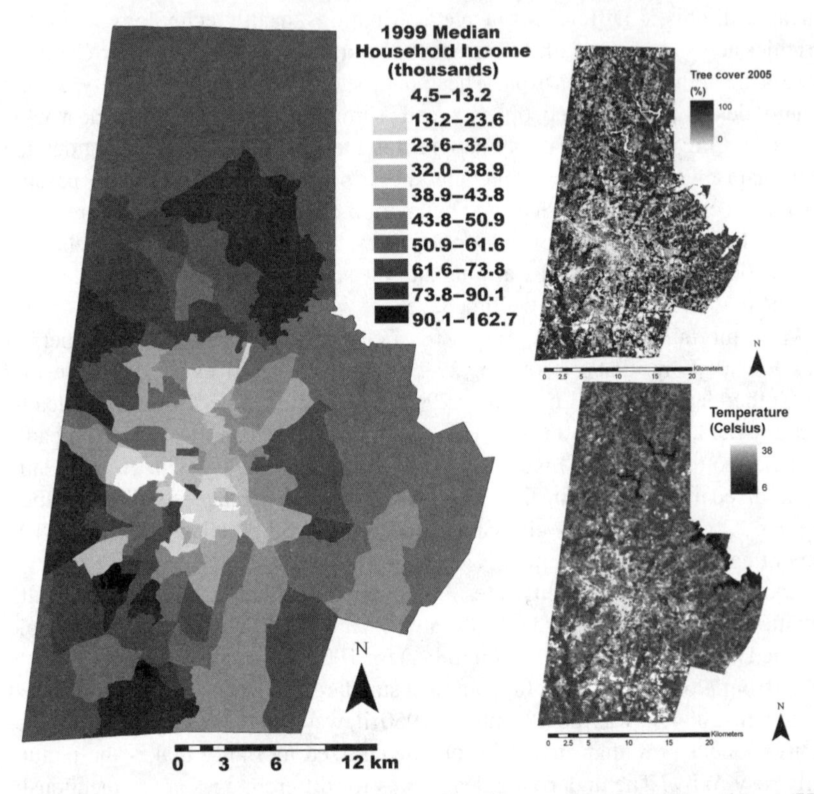

Figure 6.8: At left, Durham County's median income levels averaged over Year 2000 U.S. Census block groups (by Rob Schick). On the right, for easy comparison, are the canopy fraction (top right) and temperature profile (bottom right), previously shown and discussed in Figure 2.1. Visually, a positive correlation appears between median income and canopy cover, and a negative correlation between median income and temperature. Figure 6.13 quantifies this correlation.

The remainder of this chapter examines the benefits and distribution of urban vegetation. The images in Figure 6.8 display Year 2000 median income levels in Durham, North Carolina, averaged across census blocks, the large image at left, along with the previously discussed tree cover (top right) and temperature profiles (bottom right) across the county (see Figure 2.1). Light areas indicate low income, low tree cover, and high temperature, respectively. Visual inspection, at least by my eye, shows that all three images reflect one another, meaning that the different variables correlate with one another. For example, the lightly shaded lower-income area, just to the lower left of the city center, shows up clearly in the other two images as a high temperature and low canopy cover region. Similarly, a dark area in the northern part of the county, indicating high income, corresponds to a particularly high-income development near a watershed area having high canopy cover and low temperature. Indeed, statistical analysis presented later in this chapter (see Figure 6.13) bears out these apparent correlations: With wealth comes extensive tree canopy cover and lower temperatures.

I'm not picking on Durham: I love Durham. As I show throughout this chapter, however, environmental inequities exist throughout the United States and the world. Income–vegetation–temperature correlations materialize in many different ways across many urban areas. For example, canopy cover maps in Denver, Colorado, showed that both commercial land use and high population density correlate with low levels of vegetation.[31] Those correlations make simple common sense. To make a new building, you first have to remove grass and trees. Beyond these factors, other land-use types had complicated connections to vegetation. For example, neighborhood age generally implies higher vegetation, but features associated with small lots negate this tendency. Economically disadvantaged neighborhoods found near commercial areas had low vegetation. Sparsely populated areas with large lots owned by wealthy white people had particularly high correlations with vegetation.

I live in an area just north of the city of Durham, where, seemingly, zoning laws forbid commercial and retail activities, perhaps as a way to inhibit sprawl, or perhaps because people prefer to live far from retail spaces if they can afford it (see Figure 5.3).[32] As a result, these areas have fewer impervious parking lots, larger residential lots, lower temperatures, and fewer economic or open space disadvantages. Of course, the trade-off means that residents must drive to the nearest store to buy that emergency gallon of milk. The real need for urban open space falls right where the economically disadvantaged folks live, where, surprise, there's typically little open space.

Wealth, homeownership, and trees connect in Milwaukee, Wisconsin.

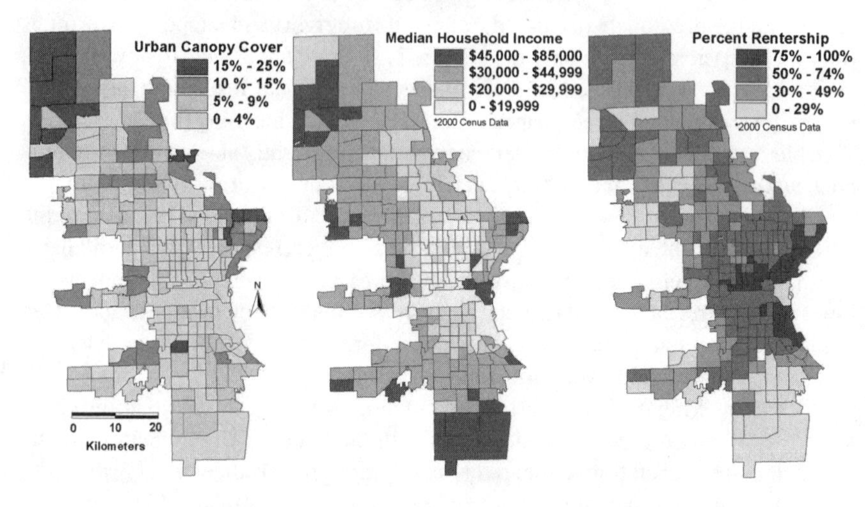

Variable	Urban Canopy	Residential Canopy
Household Income	***0.285***	***0.436***
% Rentership	-0.124	**-0.473**
% Vacancy	***-0.195***	***-0.434***
% Non-Hispanic white	***0.138***	**0.133**
% African American	-0.059	-0.013
% Hispanic	**-0.149**	**-0.250**

Figure 6.9: Maps showing urban canopy cover, median household income, and percent rentership in Milwaukee, Wisconsin (courtesy of N. Heynen, after Heynen et al. 2006). Bold numbers indicate significant correlations ($p < 0.05$), with italics even more ($p < 0.01$). Urban canopy averages tree coverage over census blocks encompassing broad land-use types, whereas residential canopy only includes places where people live.

As I mentioned earlier, Durham's not the only place where trees follow wealth. Over the next few pages, I show that clear inequities exist within cities among various socioeconomic categories, the amount of urban forests and vegetation, and healthy environmental conditions. The three maps in Figure 6.2 show the urban canopy coverage, median income, and rentership levels across Milwaukee, Wisconsin. Household income provides a strong positive correlation to both urban canopy (citywide tree coverage) and residential canopy (specific to residential land use), while rentership level correlates just as strongly, but negatively, to residential canopy cover. For example, the central portion has high rentership, low canopy, and low income. I doubt that the correlation arises because low-income citizens chop down residential trees to heat their homes, or that money grows on trees, providing the income that supports good education. Also note exceptions to the correlation, such as high rentership areas with high income and low canopy cover: These are signs of a trendier urban lifestyle.

Milwaukee has a total urban canopy cover of about 7.1%, and the city of Milwaukee manages only 4.3% of city's urban canopy cover.[33] Milwaukee's urban forester once stated that about 99% of the city lands that could have trees already have trees.[34] In other words, more than 95% of Milwaukee's urban canopy extends beyond the city's responsibility. In contrast, private residential property holds 27.8% of the tree canopy, and another 27.1% sits on commercial and industrial property. Public schools, libraries, and such have 12.7% of the canopy cover, and parks, presumably with minimally managed trees, make up about 28.1%.

These numbers indicate that increasing urban canopy has to be promoted on residential rental and commercial property, which means programs designed to enhance the property of profit-motivated landowners, who, in fact, might not associate "increased vegetation" with "enhanced property."[35] Something as simple as a free tree program designed to regreen a city, as bizarre as it sounds, adds to the inequity problem. In one example from Milwaukee, homeowners represented 89% of the people participating in the giveaway, yet that group makes up only 45% of the citizens.

In low-income areas of Milwaukee, trees mostly grow along fencelines and very near house foundations; in these situations trees are seen as nuisances that break foundations and fences as they grow too large. Removal of these trees exceeds intentional planting by others in the same urban areas.

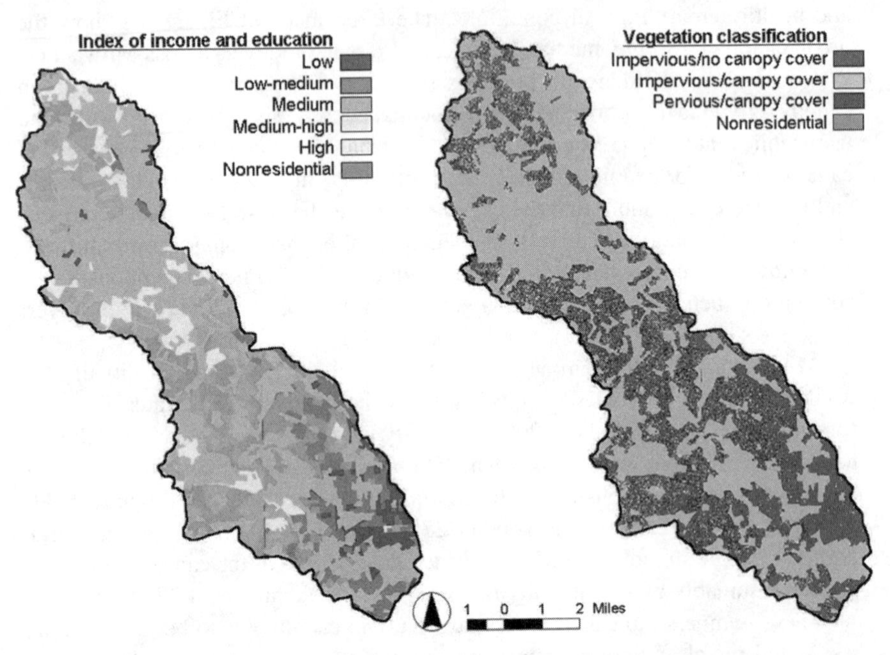

Figure 6.10: Visual correlation between vegetation and a combined measure of income and education in Baltimore, Maryland (courtesy of M. Grove, after Grove and Burch 1997). Darker areas represent lower income/education and lower levels of vegetation.

Another example connecting vegetation with socioeconomic factors comes from Baltimore, Maryland (see Figure 6.10). These plots show the correlation between lots of vegetation and high income and education; conversely, low-income areas with low levels of education tend to have low levels of vegetation.[36] Recent research on Baltimore's distribution of vegetation finds that population density and socioeconomic status, though important, incompletely predict variations in landcover. According to more recent, refined studies, understanding the connections involved in urban correlations with vegetation extend beyond taking a snapshot of a city: In Baltimore, for example, vegetative cover in the 1990s had a better connection to socioeconomic status in 1970 than it did to socioeconomic status in 1990! In other words, like the time lag between heat and mortality (Figure 6.1), the landscaping decisions made by residents 20 years ago affect vegetation patterns today more than the people who live there now.

Yet, vegetation distributions involve socioeconomic factors. Lifestyle behavior and characteristics, household lifestage, and median housing age are all important. Lifestyle behavior — a term arising from analyzing and categorizing census data — includes the social pressure to meet community expectations concerning landscaping. Household lifestage and lifestyle characteristics include such details as family composition, type of housing one lives in, and how long the members of the household have been together. Housing age also connects to vegetation status: In Durham's newest subdivisions, developers bulldoze trees (not to mention all the associated animal kingdoms) to simplify the house-building process. Thus, at house age zero there are no trees, except for the small planted ones, and certainly no frogs or salamanders. Vegetation levels can only increase!

Positive signs exist, though. In Baltimore, socioeconomic status didn't predict public provisioning of vegetation in public right-of-ways — those spaces between streets and sidewalks, parks, and other spaces under urban foresters' purview. Baltimore invokes fair practices across income and social levels, at least in this respect. However, concerning private residential vegetation, social stratification best predicted the space available for vegetation — parcel area minus house area — while lifestyle behavior best predicted whether that space has vegetation. Disproportionate rentership, tied to low education and low-income areas, likely explains the correlation's source: Tenants have little motivation to landscape their landlord's property.

Parks, trees, and plants come with wealth.

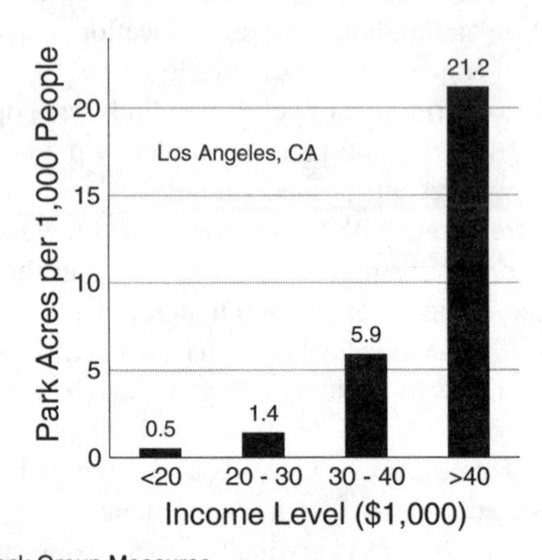

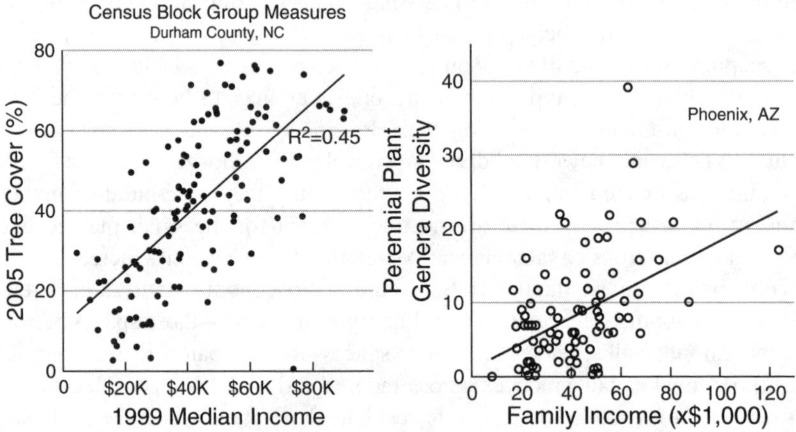

Figure 6.11: Top plot shows park provisioning in Los Angeles against 1990 median household income (after Wolch et al. 2005). Higher income groups have 40 times more park area per person. Bottom left graph shows the dependence of tree cover and income in Durham County, North Carolina, using the data in Figure 6.8. The bottom right plot shows that plant diversity increases with median family income around Phoenix, Arizona (after Hope et al. 2003).

Let's put some of these visual inequity correlations into graphical form with results compiled from across the United States. At the top of Figure 6.11, results for Los Angeles demonstrate that wealthier people have access to more park acres per person.[37] The team producing these results also demonstrates that racial inequities exist in park provisioning (see Figure 6.12) and that park funding sometimes heightens the inequities when, for example, considered on a per-child basis. Resolving these inequities won't be easy or cheap, but they might involve retrofitting vacant properties, public properties such as schools and libraries, and riverbeds.

At bottom left, I've also plotted data from Figure 6.8 showing how Durham County's tree coverage depends on median income. These points plot the tree canopy against the median income averaged over the 129 census block groups from the 2000 Census for Durham County. The two variables correlate extremely strongly[38] with income explaining 45% of canopy coverage. Astonishingly, each additional $10K in median income means that a neighborhood has an additional 7% of its area covered by trees. These results demonstrate that Durham's poorest citizens have less than 20% tree coverage, while its wealthiest residents have 70–80% tree coverage. However, note the outlier at an income level of about $72,000 and no canopy cover. That point could be either a newly cleared subdivision, like that pictured in Figure 2.17, or situated on a golf course with few trees. Of course, the correlation doesn't reveal mechanism: Perhaps low-income areas have their trees chopped down and replaced with impervious surfaces, or, more likely, wealthier people choose to live where there are trees.

One last example, at bottom right, demonstrates an example of provisioning biodiversity, measured in terms of the number of plant genera ("genus" is the botanical classification above species) versus family income in Phoenix, Arizona.[39] This study examined 204 30 m-by-30 m plots across a variety of land-use classifications, including urban, desert, agricultural, and transportation. Although plant genera richness correlated significantly with family income, urban land use and elevation showed even stronger correlations. Previously, I showed results indicating that measures of well-being were heightened in parks with greater biodiversity; here we see inequitable distributions of biodiversity. Importantly, surveys of plant biodiversity have included horticultural species, and one of the mechanisms for maintaining the correlation with income arises through the maintenance cost of exotic species, leading to a phenomenon the researchers called the "luxury effect."

Minority populations have worse air, income, and asthma.

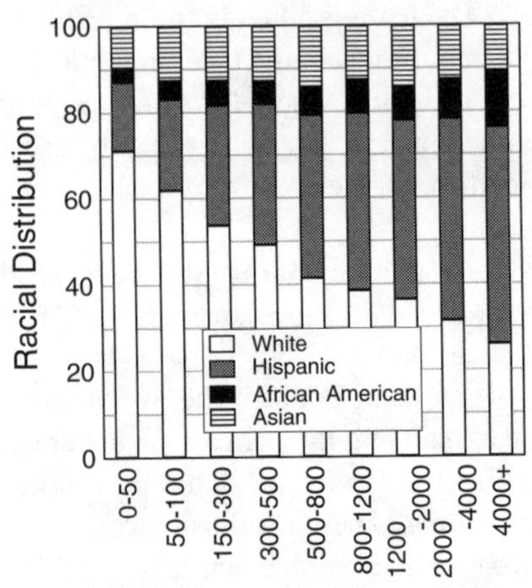

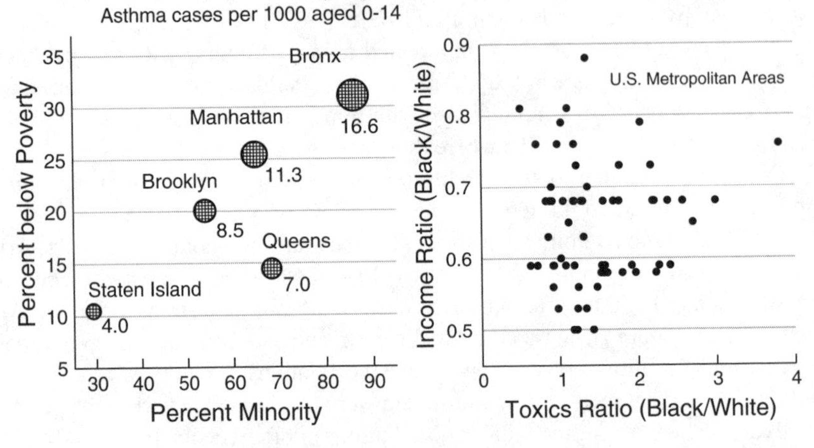

Figure 6.12: Top graph shows park provisioning in Los Angeles (after Sister et al. 2007). Bottom left graph shows that asthma hospitalizations differ between the boroughs of New York City, increasing with percentage of minority population and poverty level (after Maantay 2007). Circle area scales with the number of cases. Right plot shows the toxics and income ratios of blacks to non-Hispanic whites for 61 metropolitan areas across the United States. Independence of racial background occurs when both ratios equal one (after Downey 2007).

Indications of racial inequities in vegetation and tree canopy coverage come through in the correlations of Figure 6.9; in Figure 6.12 I show several results that make those inequities clearer. At top I show the racial distribution of park provisioning in Los Angeles. This study, associated with the Los Angeles study in Figure 6.11, determined the number of people sharing each park available to citizens.[40] Essentially, think of the number of people per park as the number of people for which that park is their closest one. When parks are abundant, or population densities are low, a park has very few people, and when parks are far and few between, or population densities are very high, a park has many people. In Los Angeles, parks that service relatively few people have racial distributions dominated by whites, but parks with relatively many people (100 times more than the lowest population levels) service predominantly Hispanic populations.

The bottom two plots link racial and economic condition. The bottom left graph shows the asthma hospitalization rates per 1,000 children across the boroughs of New York City, plotted against each borough's poverty rate and percentage minority population.[41] Extending this observation of inequity nationwide, a study of the 61 largest metropolitan areas (areas of more than 1 million people) across the continental United States is shown in the bottom right graph.[42] These data rely, in part, on the "segregation index" — an important statistic that defines the fraction of the minority population that would need to move to achieve an even distribution across the metropolitan area. Over these cities, in the year 2000, the segregation index varied from 84 down to 37. With this statistic, the ratio of the median household income for blacks and whites spanning these metropolitan areas varies from a low of 0.5. This means that whites have twice the household income, on average, as blacks in the most segregated cities, up to the highest income ratio of 0.88 in the most integrated cities. Another inequity measure, the "toxics value," compares different kinds of air pollution in different locations by weighting each pollutant by its toxicity. The "toxics ratio" defines the relative toxics value in black and white neighborhoods in these metropolitan areas. The bottom right plot shows the racial inequity of income and air pollution, which reflects the results found for tree cover and racial and socioeconomic correlations.

The Raleigh–Durham–Chapel Hill area has a segregation value of 46.2, a black-to-white income ratio of 0.59, and a black-to-white toxics ratio of 1.54. These values put this area in a roughly average situation. It seems reasonable to assume that a similar inequity exists in asthma hospitalizations and park provisioning, for example, but that would only be a guess.

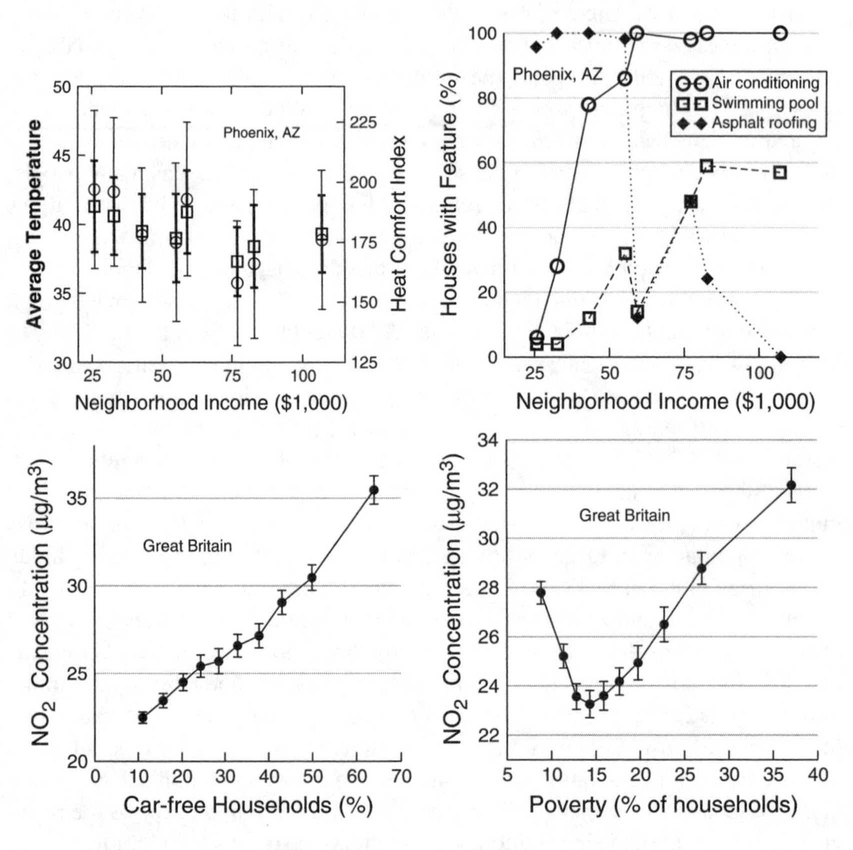

Figure 6.13: Top left graph shows that a neighborhood's average temperature and heat comfort index (lower values indicate more comfort), and at top right, various "cool home" features, depend on average income in Phoenix, Arizona (data from Harlan et al. 2006). Error bars reflect standard deviations; both the temperature and heat comfort indexes decrease significantly with income. Averaged over all wards in Great Britain, the bottom plots show that NO_2 pollution is higher, remarkably, in wards with lower car ownership (at left) and in wards with higher poverty (after Mitchell and Dorling 2003).

Better environments tend to be found in wealthier areas, but people's choices complicate the patterns. Matching expectations, the top graphs in Figure 6.13 report temperatures (measured at 5 PM during the summer of 2003) were higher in poorer neighborhoods, as well as the heat comfort level, in Phoenix, Arizona, where this study was performed.[43] Although the error bars look large, they're standard deviations,[44] the temperatures had significant differences and the dependence of heat comfort on income was also significant. Researchers also found that vegetation was significantly correlated with the heat comfort index, and we've already seen one case of vegetation being correlated with income in Figure 6.10. One question with these results, perhaps, is that, although the correlations with income are significant, are the correlations important? Maybe everyone in Phoenix is uncomfortably hot, but those with high income are a little less uncomfortable: Is there a cool spot in the oven? The importance, in this sense, isn't completely clear.

Another important point is that in a place like Phoenix, where water is in short supply, having lots of vegetation must be considered carefully because of the water demands. In Durham, North Carolina, we get plenty of rainfall — except for the occasional, exceptional drought and seasonal summer dryness — so water isn't usually a limiting problem for promoting vegetation.

Results from the Great Britain study on NO_2 correlations break the stereotyped expectations. Pollution levels decreased with greater car ownership, at bottom left, because urban areas have higher pollution levels, along with a reduced need for car ownership if you live there.[45] Some neighborhoods have high pollution levels, but these people own few cars and have low emissions. People from outlying areas drive their cars to these urban areas, and the urban dwellers have to breathe the result. In some sense, this situation illustrates an environmental injustice because these people breathe pollution not of their own making, but one moderates one's sympathies because these folks have chosen a trendy urban lifestyle over cleaner air in wealthy suburban neighborhoods. This choice also drives the higher pollution levels at low poverty levels at bottom right. The real environmental injustice occurs because some of the people in some of these urban areas experience greater poverty than average, breathe others' pollution, but can't afford to move to healthier environments.

Talking about equity, health, and vegetation presents a complicated picture. Another study, this time of England alone, found that vegetation promoted good health in urban areas and for the rural poor, but not for the rural wealthy and the suburban poor. Rural wealthy residents don't rely on public greenspace provisioning. Morever, the study found that poor suburban residents have generally worse health and lower quality greenspace, but specific mechanisms remain unclear.[46]

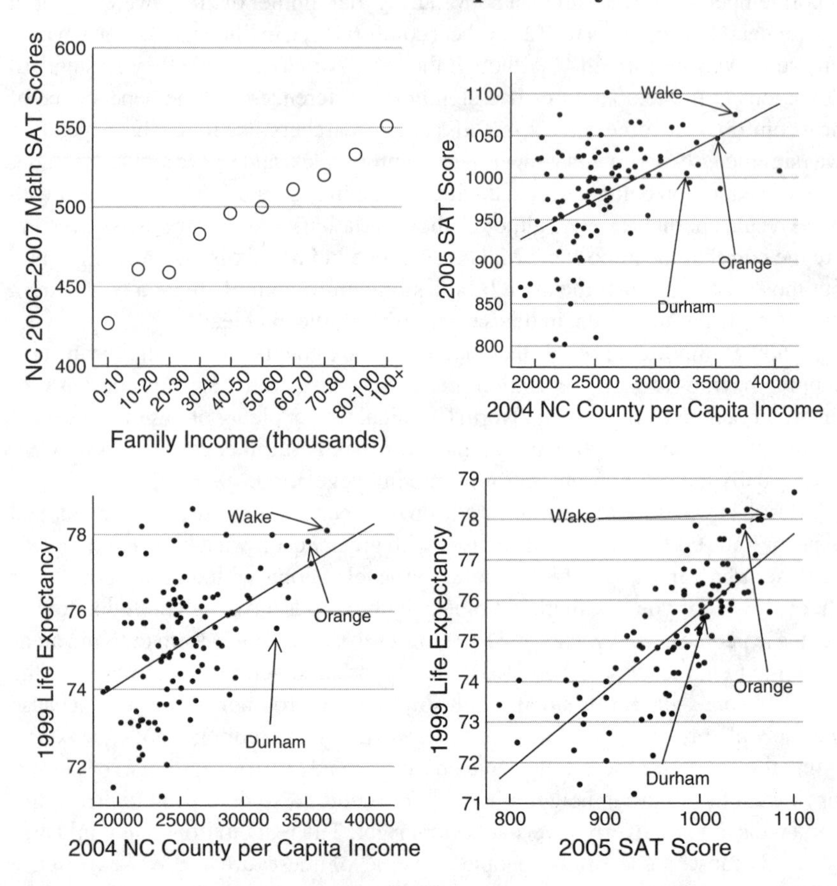

Figure 6.14: Top left graph shows how the North Carolina math SAT scores from the 2006–2007 academic year vary with self-reported family income (data from The College Board 2007 NC statistics). Top right graph plots the 2005 combined SAT scores against the 2004 per capita income for each county in North Carolina (data from www.ncpublicschools.org). Bottom plots show, for each North Carolina county, life expectancy (from Ezzati et al. 2008) versus income and combined SAT scores. Every $10,000 in per capita income or additional 100 SAT points yields about two additional years.

Again, let's put urban nature into the proper context, as was done with the heat-related mortality of older people. Urban vegetation helps children develop positively, as previous results demonstrate, but are other factors more important? At top left of Figure 6.14, for example, are the Scholastic Aptitude Test (SAT) mathematics scores for the 2006–2007 academic year for North Carolina students. Students themselves reported these incomes,[47] and assuming the reported incomes aren't too far off, family income shows a strong correlation with this developmental measure.[48] At top right, county-level data from North Carolina — 2005 combined SAT scores vs. 2004 per capita income — confirm a strong dependence on family income.[49] Certainly, some counties with low incomes perform very well, but students in counties with higher per capita incomes consistently perform better. As a result, for every $10,000 increase in a county's per capita income, the county's SAT scores increase by nearly 75 points.

Another set of correlations shows how county-level life expectancy increases with county-level per capita income and SAT score.[50] As plotted against per capita county income at bottom left, life expectancy mirrors SAT scores.[51] With this correlation, the data imply that every $10,000 in per capita income yields two additional years of life expectancy. In addition, the correlation between SAT scores and family income implies the obvious correlation between life expectancy and SAT score depicted here at bottom right.[52] According to the correlation, an additional 100 points on an SAT score implies an additional 1.9 years of life expectancy.

Remember, these plots show correlations, not causation or mechanism, and the two correlations with life expectancy should be expected because of the correlation between income and SAT score. Mechanistically, perhaps people with higher incomes can afford better health care. Perhaps people with higher SAT scores know how to live in a more healthful manner. Perhaps some other aspect of people's lives underpins both these correlations. Precisely because these correlations exist, scientists should have the resources to uncover their underlying causes.

So, is urban nature beneficial to childhood development? Figure 5.10 showed that effect, so the answer is yes; further, I suspect that effect carries across different cities. But is it a strong effect? Assuming SAT scores correctly measure well-developed children, many factors beyond vegetation, associated with family income, help children develop positively.

Chapter 7

Summary and Implications

I've covered many factors that connect to urban environmental issues, including energy and carbon, air, heat, water, nature, and the health and welfare of people, as well as the effect cities have on regional and global climatic patterns and the changes imposed on other species. Cities, by their very definition, involve natural lands turned into rain-repelling, impervious surfaces — roofs and parking lots — filled with people and cars. As cities become bigger and denser, impervious surfaces increase while trees, forests, parks, and grass decrease.

People in cities use an unusually large amount of energy and water in limited spaces, and the resulting problems include air and water pollution, heat and health problems, and reduced benefits from natural ecosystems. Cities create many environmental problems.

Despite these problems, an economy of scale provides an effective argument for bigger, denser cities. Urban states have lower per capita energy use than rural states: With high densities come reduced individual transportation demands and lower electrical transmission line losses. Extending this argument, if vegetation and forest remnants serve no function in urban settings, cities should increase densities as high as possible, creating square kilometers of contiguous multistory enclosures with internal moving sidewalks transporting the people within. At this limit, cities would become vast "underground" bunkers — land-based cruise ships — with small portals for food provided by rural areas and air exchange for oxygen provided by external forests. Their green coverings could collect all their water from rainfall and all the electrical power they need from solar energy, with, perhaps a baseball and soccer field here and there.

At the other extreme, people could spread out as much as possible while attempting to reduce urban environmental issues associated with aggregating impervious surfaces. Unfortunately, there's not much room to spread. Suppose Durham County, North Carolina, citizens were removed from cities and dispersed evenly across the county. The density would be 1.35 people per acre. Suppose North Carolinians were removed from cities and dispersed evenly across the state. The density would be one person per 3 acres. For all Americans there'd be one person per 8 acres. For all humans, there'd be about 5 acres per person. Ignoring deserts and counting only the world's arable land, the 10% of the land surface where crops can grow, that number drops to half an acre per person. We have such a large human population that we can't live without cities.

Another possibility is to imagine a local region divided into three parts: agricultural; forested watersheds with no impervious surfaces; and cities built out to 100% imperviousness and maximum population density. In this situation the issue becomes that of where a small forest parcel would be best placed: within the city as a small urban patch, or as part of an imperceptibly larger contiguous forest at the city's edge. Placement also depends on the anticipated function of that forest fragment, the function or functions that defines the term *best* in the previous sentence.

People in developed countries use large amounts of energy. Of course, using a lot of energy is, today, one mark of a developed country, and the challenge of the next few generations involves reducing energy use or finding a sustainable or inexhaustible energy source without environmental hazards. The direct problem, again, reflects our large population and high energy use, with too many vehicles and too little mass transit. This transportation situation demands too many impervious surfaces — roads and parking lots — leading to too much driving, too much urban heat, and too much air pollution. The urban heat island directly results from large expanses of impervious surfaces and the concomitant lack of vegetative cover, though it is somewhat moderated by city architecture. Sadly, urban nature plays no consequential role in offsetting our fossil-fuel use through carbon sequestration, nor does it provide the all-encompassing energy source for America's energy-rich lifestyles. Even a naturally vegetated and sustainably harvested United States can't satisfy that demand. Americans simply use too much fossil fuel to hope that planting a tree in everyone's backyard will address the carbon and energy problems.

Vegetation and trees certainly shade impervious surfaces by absorbing and reflecting light, and cool the air by evaporating water. These natural solutions cool cities. Other solutions include painting impervious surfaces white, reducing their heat absorption; in fact, compared to trees, tall grass does a much better job at cooling through both shading the ground and evaporating water. Replacing impervious asphalt with grassy paving provides both types of cooling, allowing some stormwater infiltration while also providing parking. At least in areas where cooling takes precedence over heating, strategically placed deciduous trees can make a difference, but with ample water, long grass might do a better job.

A well-placed tree, one situated, say, in an urban canyon, efficiently and beautifully interrupts the exchange of radiant energy between thermally massive structures. But these spots tend to be in high-traffic areas that stress even the toughest trees and demand constant maintenance. Near homes, a well-placed tree might have some hopes of reducing energy use, yet, more often than not, the energy reductions promoted in summer cooling simply offset energy increases in winter heating, and the use of fossil fuels in pruning, maintenance, and careful removal when dying trees are too close to buildings adds costs. A well-insulated house minimizes much of the potential benefit, not to mention having a cold house in the winter and a warm house in the summer.

Coal-burning power plants and petroleum-powered traffic add reactive nitrogen to urban air, providing a critical link between cities and harmful air pollution. In particular, I presented the steps that lead to high ozone concentrations, but many other pollutants, including the nitrogen itself, cause problems.

Vegetation reduces these pollution levels through two paths: first by slowing down air as it moves through branches and leaves, letting heavy particles fall out; and second by absorbing pollutants into their tissues through open stoma. Well-chosen urban trees in the right spot can provide some pollution-reduction benefits. But can we consider vegetation as pollution scrubbers any more than we can the lungs of people and animals as they breathe city air? All organisms sequester many of these toxic chemicals and suffer adverse consequences, but do organisms efficiently remove from the atmosphere what came out of our automobiles' tailpipes? Empirical studies just don't support large increases in air quality due to trees, other than displacing impervious surfaces and reducing traffic. Similar to our carbon problem, directly reducing emissions from burning fossil fuels likely provides the best strategy. Effective legislation has produced cleaner air over the last few decades, and further regulations demanding cleaner energy sources seemingly make the best sense.

We know that even very small amounts of impervious surface in a watershed harms water quality, changes the fundamental nature of urban streams, kills aquatic organisms, reduces the pollution-sequestering abilities of these streams, and impairs our drinking water sources. We've seen that impervious surfaces connected directly to stormwater systems, which flow directly into streams, are the biggest water quality problem. Yes, farmers use fertilizers and produce concentrated animal wastes that sometimes run off into streams. They pay for fertilizers, and their economic constraints and task of feeding the immense human population demand that they use it. But suburban residents nurture lawns, fertilizing them simply out of habit with little functional purpose other than a ritualized, nonmarket pleasure. Along with atmospheric nitrogen falling on impervious surfaces, lawn fertilizers get washed away with stormwater into nearby streams. Construction sites cause problems, too, when grading and digging remove vegetation and living soils, exposing easily eroded dirt that washes into streams, burying bottom-dwelling organisms and ecosystems with sediment.

Improving water quality means filtering out molecules and sediments from every gallon of rainwater, an important job that forests perform naturally in rural areas, and open space can perform in urban areas. An important goal for cities could be to make certain that every stormwater drains empties into buffers and constructed wetlands without overwhelming their biological limitations. Relatively intact urban forests, soils, and constructed wetland ecosystems may be the most maintenance-free approach to handle the large water volumes involved in stormwater flows.

An interesting complication surrounding Durham's stormwater situation, certainly repeated across the country in many jurisdictions, concerns the issue of scale. Durham's water comes from reservoirs in the northern parts of the county, filled by more rural and forested areas, lacking many sediment and pollution problems. Some critical watershed lands have been purchased with county and state taxpayer money for that express purpose. However, the City of Durham's stormwater drains into the Falls Lake and Jordan Lake reservoirs, with that water destined for drinking by citizens of other municipalities. Wake County is one of those municipalities, and it has a much higher household income. One can legitimately argue whether cash-strapped Durham citizens ought to spend money on constructed wetlands and urban open space to improve Wake County's water quality, taking on changes in land use that might even reduce Durham's tax base at a time when Durham County schools need more resources.

Trees and natural vegetation provide the right environment for positive social interactions, calm domestic stresses, and enhance child development. Wealthy residents can afford the choice to live in treed or untreed neighborhoods, and correlations of vegetation and income bear evidence of the most commonly preferred choice. In those places, citizens rally against developments with impervious surface and call for public funding of natural open space. Seemingly, these are not the citizens with the greatest need for the social benefits from publicly funded trees.

Low-income citizens live with less vegetation, perhaps without affordable options. In the places where they live, additional natural areas might serve as social gathering points for positive social opportunities and might reduce crime rates through enhanced community vigilance.

Perhaps the most important and lasting benefit of vegetation involves better developed children with a higher sense of delayed gratification. The ability to sacrifice short-term pleasures for long-term benefits has implications for every aspect of life: The important lessons behind planting a seed in April and harvesting food in July can't take place in parking lots. I suspect similar lessons are learned each spring, perhaps passively, from watching perennials resprout and trees leaf out. If a child's bedroom window shows only roads, parking lots, and strip malls, growing children lose this natural education. Social justice demands promoting urban open spaces, and every child should have the view of a tree outside his or her bedroom window.

Admittedly, the evidence for these positive social aspects for vegetation remains scant, and much of what evidence exists arose from studies of distressed social conditions. We need more studies across a range of socioeconomic levels pursuing the human social benefits of urban open space.

What's the dollar value of enhancing the natural environment of these children growing up in poverty? I use Durham County as an example, where the U.S. Census reports about 40,000 Durham citizens in poverty, with about 10,000 of them under 18 years of age. Suppose we could hope for a 5% gain in childhood development due to enhanced vegetation and trees, a gain on the high side of the studies performed at the Chicago public housing developments. That number fully admits that 95% of a child's development comes from more proximate causes, unrelated to backyard nature. Natural areas provide a benefit, but a small one compared to parents, friends, and teachers. Also suppose, rather pessimistically but erring on the side of underestimation, that these children become adults who live in poverty, earning about $10,000 per year. If growing up in a more natural environment resulted in a 5% "better" development, say through enhanced ability to delay gratification, and if that enhancement resulted in a 5% gain in income, then each of these citizens would earn an additional $500 each year. A small gain, yes, and perhaps realized by just a few children planting seeds for the future, lifting themselves out of poverty. Considering all of Durham's 30,000 adults who live in poverty, that increased productivity could yield an additional total income of $15 million per year to this population. Even assuming that enhanced vegetation yields only a 1% gain in childhood development, the dollar benefit might still reach $3 million per year. Certainly, these numbers justify a significant program enhancing vegetation where poor people live that would pay for itself through better lives. Enhancing urban natural areas today for better, higher-earning citizens tomorrow truly represents a city's delayed gratification.

In a nutshell, urban nature does not solve, to any appreciable extent, our looming carbon dioxide and global warming problems. I also maintain great doubts that urban trees significantly reduce air pollution, or that they provide a particularly cost-effective approach to reducing residential energy use. Seemingly, the two primary benefits of urban nature involve enhancing the social welfare of people's lives and helping alleviate the environmental problems associated with stormwater runoff. In other words, "Hold your water and watch nature."

Appendix

Graphical Intuitions

Essential Graphical Knowledge

Graphs provide tremendously important information to scientists, and in this book I rely on them heavily. Graphs encapsulate much information, providing an easy look at broad patterns in scientific data. These subtle and broad patterns of variations and trends usually remain completely hidden in tables of numbers. Words and descriptions yield incomplete yet lengthy prose, prone to misinterpretation, whereas graphs contain all of the information.

In this Appendix I point out important visual messages that graphs display, messages that only graphs can reveal concisely, messages that help people understand more deeply the phenomena surrounding urban environments.

Three equivalent data representations.

Treatment Type	White Response	Black Response
Red	1.25 (0.4)	1.31 (0.5)
Green	2.51 (0.3)	1.54 (0.4)
Blue	1.75 (0.2)	2.28 (0.6)

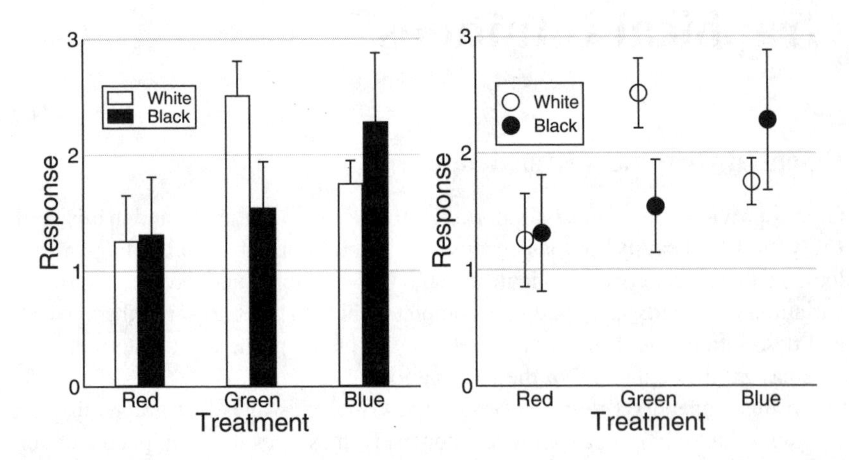

Figure A.1: A table, a bar graph, and a plot are three different, but equivalent, ways to show the same data. No one way is best in all cases, and each is best in some cases. The table displays just a few made-up numbers, along with their uncertainties in parentheses. Though "treatment" and "response" imply active experimentation, one would treat observational categories and correlated features similarly.

Scientists present data in many different ways, dependent on their desired emphasis, economy of journal space, and personal tastes. Here in Figure A.1 are three different presentations of one set of made-up data.

Tables of data hold information; this is a greatly useful feature allowing other people to process the numbers, but one's focus gets drawn toward individual numbers, not general patterns. I provide one such example in Figure 3.15.

The bar graph with the uncertainty scale sitting on top, sometimes derisively called "pinhead plots," emphasizes levels of effect across different treatments. I discuss uncertainties later on in Figure A.6; here let's just think of them as uncertainties in the listed values. Some scientific disciplines use bar graphs extensively, and I use them in part to maintain the original publication's spirit and in part because bar graphs provide good visual comparisons when treating categorical variables. Sometimes experiments consist entirely of two treatment categories, or experimental scenarios, like the status of vegetation, barren versus green, or a few treatment levels, low, medium, and high. Bar graphs simplify comparing several measurements at these values by a quick look at differing patterns (Figure 5.9 provides three good examples).

The graph with data points repeats the information in the bar graph. In this case of very few points, a bar graph and data plot provide equivalent representations. However, situations where the horizontally plotted variable, identified here as the treatment, takes on many values and the response variable on the vertical axis changes in a broad scale manner over these values, one wants to see how the line changes. Often there are so many data points that they make up a continuous sweep, and a scientist plots a line, not individual points.[1] One might draw a line between the data points, but in the situation of a categorical variable, a line wouldn't make sense. No continuous connection exists between the different categories, presumably, and the guide to the eye that the line provides has no value as an interpolating device. With a continuous variable on the x-axis, however, a line joining the data points provides assistance envisioning the full connection between the two axes' variables. For example, replace red, green, and blue with the temperatures 60F, 65F, and 70F; then extend the experiment from 40F to 100F by 1F increments and add a few shades of gray responses. Tables of numbers become data archives, and only plots, not charts, are readable.

Graphing visual correlations.

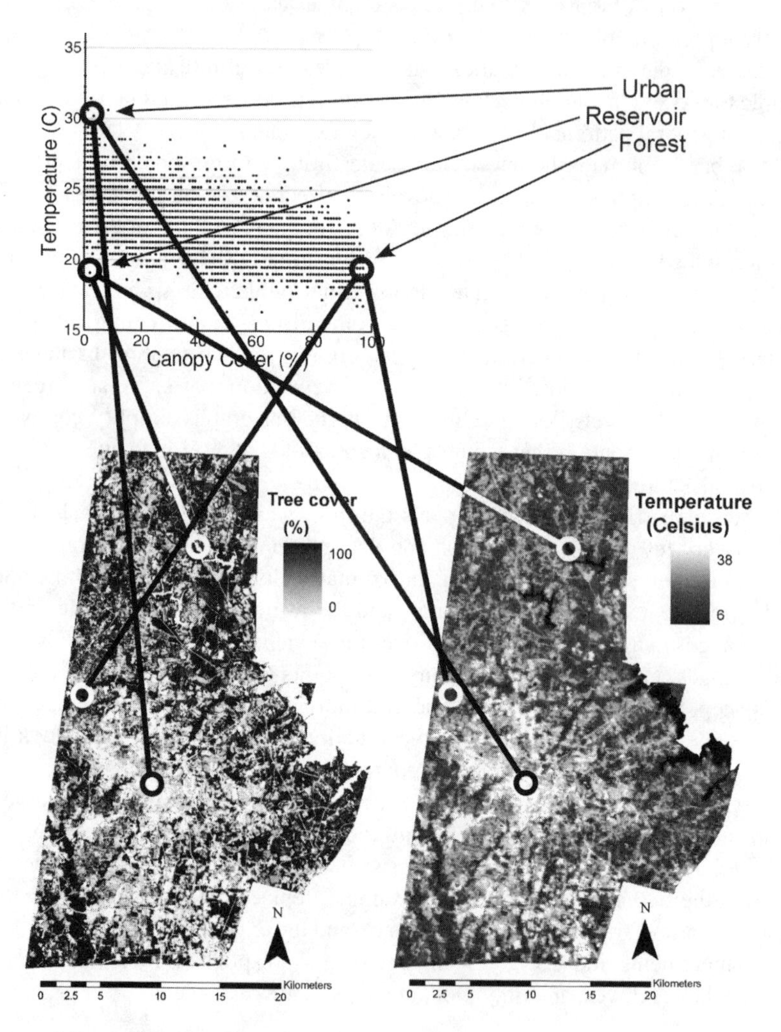

Figure A.2: These two maps show Durham County's 2005 canopy cover, bottom left, and thermal image, bottom right. My eyes show strong visual correlations between the two images, but scientific analysis requires numbers. My lines connect representative points in the images to appropriate data points in the graph.

In Figure A.2, I describe the connection between visually observed data and plotted data, and explain why scientists pursue both approaches. The graph at top plots temperature and tree canopy cover values from the images, showing that low tree cover corresponds to high temperatures, and high canopy cover to low temperatures. I point out three specific examples in the images. The urban point shows light areas in both tree cover and temperature, meaning high temperature where no trees sit. An unusual point, low temperature with low tree cover, identifies a drinking water reservoir in northern Durham County, with a light area in tree cover and dark area in temperature. Forested areas have dark areas for both tree cover and temperature, meaning a cool, tree-covered location.

Let's not worry here about the mechanisms — thermal mass, urban heat island, and all that — that join the two variables displayed in the images of Durham County's canopy cover and thermal profile. I extensively discuss the science behind these images and graph in Figures 2.1 and 2.3, and I use them here just to explain several things about graphing.

For my purposes here, just note that the two variables, for whatever reason, covary across Durham County, and the basis for our sense of covariation arises from visual inspection and pattern matching. Of course, I hope everyone sees what I see: Light areas in the left image coincide with light areas in the right image. If someone doesn't see this correspondence, it just emphasizes that science doesn't get very far by visual inspection alone because visual inspection varies from person to person in a nonquantifiable way. Indeed, color-blind scientists may have a particularly tough time with arbitrarily chosen color palettes. We need to quantify the patterns and apply some statistical analyses to answer a specific question like the extent to which variation in one variable explains the variation in another, answered by a parameter like R^2 discussed in Figure A.3.

However, the upper plot is incomplete. The images reveal one feature not shown in the upper plot. Notice how pixels with low tree cover (white) and high temperature (white) clump together in the county's central area, where the City of Durham sits. This clumpiness defines a spatial correlation. Imagine taking all the cover–temperature pixel pairs and shuffling them all throughout the county, making sure pixel pairs are shuffled to the same location. The county images would take on an even, randomized gray tone, but the graph at top wouldn't change at all. Unshuffled, the images clearly show more information than the top plot.

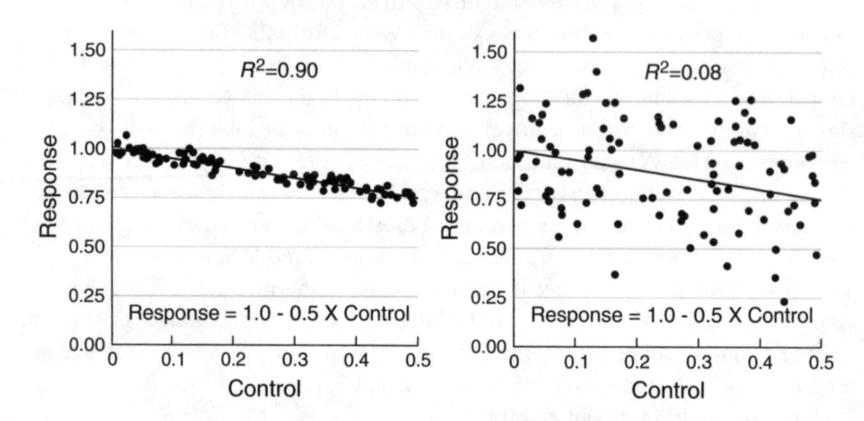

Figure A.3: These two graphs distinguish the difference between a significant correlation and an important correlation. Both graphs show significant correlations between control and response, but the plot at left has a really *important* (meaning strong) correlation, while the one at right does not. The statistical quantity, R^2, measures importance, translating roughly to the fraction that the control parameter contributes to determining values of the response variable. Thus, $R^2 = 0.9$ and $R^2 = 0.08$ mean that control parameter variation "predicts" response values, in the left and right plots, respectively, with 90% and 8% strength.

The graphs in Figure A.3 reveal the meaning of an "important" correlation. Though I don't show specifics of the statistical measure called R^2 — details can be found in any basic statistics text or website — both graphs have a significant correlation between the control parameter and the response variable.[2] I know this correlation exists because I made up the data with a computer program[3] based on the displayed formula, precisely with the intent of describing statistical importance. I know, beyond any doubt, that a significant correlation exists between the two variables. Linear regression, the procedure determining the best-fit straight line to control–response data like these, would produce an expression very close to that mathematical equation (within statistical bounds). Whether or not a regression, or statistical connection, demonstrates "significance," or exists beyond reasonable doubt, comes from statistical considerations I won't cover here. That said, many tests of significance state a value for the variable p, and $p < 0.05$ typically denotes a significant effect.[4] Essentially, that value means that the observed connection would happen by chance just once out of 20 times ($0.05 = 1/20$).

At left I show a plot that has a really *important* correlation between the control parameter and the response variable. A statistical quantity, R^2, measures importance, and in this case, $R^2 = 0.9$ means that control parameter variation "determines" 90% of response variable variation. In other words, if you know the control parameter value, you have 90% of the information needed to predict the response value. If a correlation with this strength popped up in, say, some type of health problem and an environmental variable, you can bet that scientists (not to mention medical doctors and public policy folks) would sit up and take notice.

However, in the plot at right, with the same linear dependence between control and response features, the response variable has a tremendously greater amount of variation completely unrelated to the control parameter's value. For example, near a control value of 0.15, the response variable ranges between extreme values of 0.4 and 1.5. In this case, the correlation remains significant, but its importance measure drops to just $R^2 = 0.08$, meaning control parameter variation predicts only 8% of the response variable's variation. The "control" just isn't as important to the response, leaving 92% of the response variable's variation unpredictable.

As a real comparison of importance versus significance, compare Figures 4.13 and 5.10.

Plotting transformed data.

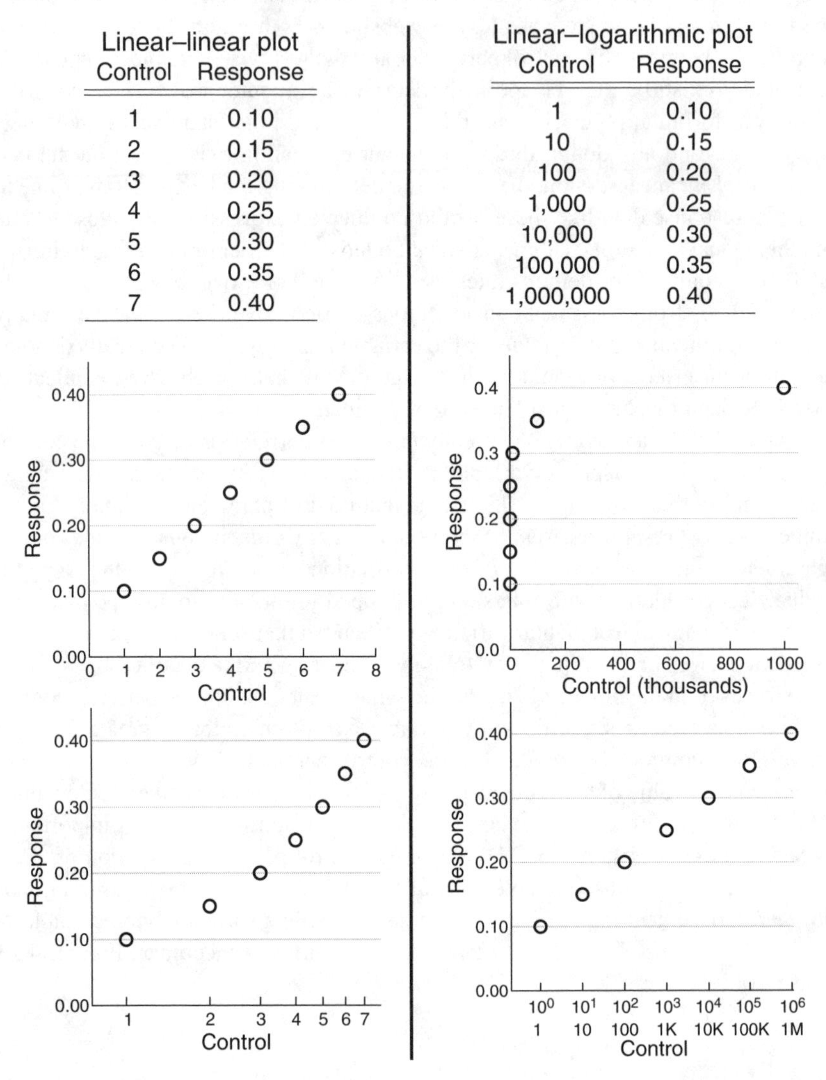

Linear–linear plot

Control	Response
1	0.10
2	0.15
3	0.20
4	0.25
5	0.30
6	0.35
7	0.40

Linear–logarithmic plot

Control	Response
1	0.10
10	0.15
100	0.20
1,000	0.25
10,000	0.30
100,000	0.35
1,000,000	0.40

Figure A.4: Tables of data appropriate to linear and logarithmic plotting axes on the left and right, respectively.

The idea of a logarithmic axis confuses many students reading graphs. In some cases, like the one I made up in the left-hand side table in Figure A.4, the X and Y variables — control and response — relate to one another in a linearly dependent manner: Increasing the control from 1 to 2 increases the response by 0.05; increasing the control from 6 to 7 also increases the response by 0.05. We see this directly proportionate response by the straight-line, constant slope in the upper left graph. Not surprisingly, we say that the response has a linear dependence on the control parameter. Any other type of response would be called not linear, or "nonlinear." Most variables depend on their underlying factors through mechanistic processes that lead to nonlinear responses.

Normal, untransformed plots rely on the connection between X and Y embodied in the equation for a straight line, $Y = mX + b$, where m is the slope, and b is the Y-intercept, the value of Y when $X = 0$. The slope tells how much Y changes when X is changed by one unit of value, like the 0.05 described above.

Other situations exist, for example, $Y = cX^d$, $Y = cd^X$, and $X = cY^d$. In these situations one takes the logarithm of both sides of the equation, yielding $\log Y = \log c + d \log X$, $\log Y = \log c + X \log d$, and $\log X = \log c + d \log Y$, respectively. Thus, depending on the relationship between the variables, the best axes transformations follow.

I depict one special nonlinear situation in the table on the right-hand side. Here, compared to the linear example, the Y variable depends more weakly on the X variable, and the response saturates as the X variable increases. In this imaginary case, increasing the control parameter by one order of magnitude, from 1 to 10 (an increase of 9), increases the response by 0.05, while increasing the control from 100,000 to 1,000,000 (an increase of 900,000) also increases the response by 0.05. The effect on the response variable of a certain increase in control depends on the control value itself. That differential response wonderfully describes a nonlinear response.

When one variable spans several orders of magnitude while the other doesn't, one plots the graph using a linear axis for the constrained value and a logarithmic axis for the widely varying one. I first show, in the upper right graph, what happens if you don't choose the correct plotting option and instead use two linear axes. All the X values bunch up near the origin, making the dependence extremely difficult to see. These bunched-up values need to be stretched out, and the lower right plot achieves exactly that goal. In fact, looking at the tick labels along the bottom, we see that the exponents, for example the "4" on 10^4, written above the 10K (meaning 10,000), have a linear progression on the logarithmic axis. This is exactly as it should be: We can write any number as $X=10^Z$, in which case, $\log X = Z \log 10 = Z$. The logarithmic axis plots the exponent linearly.

Quantities covary.

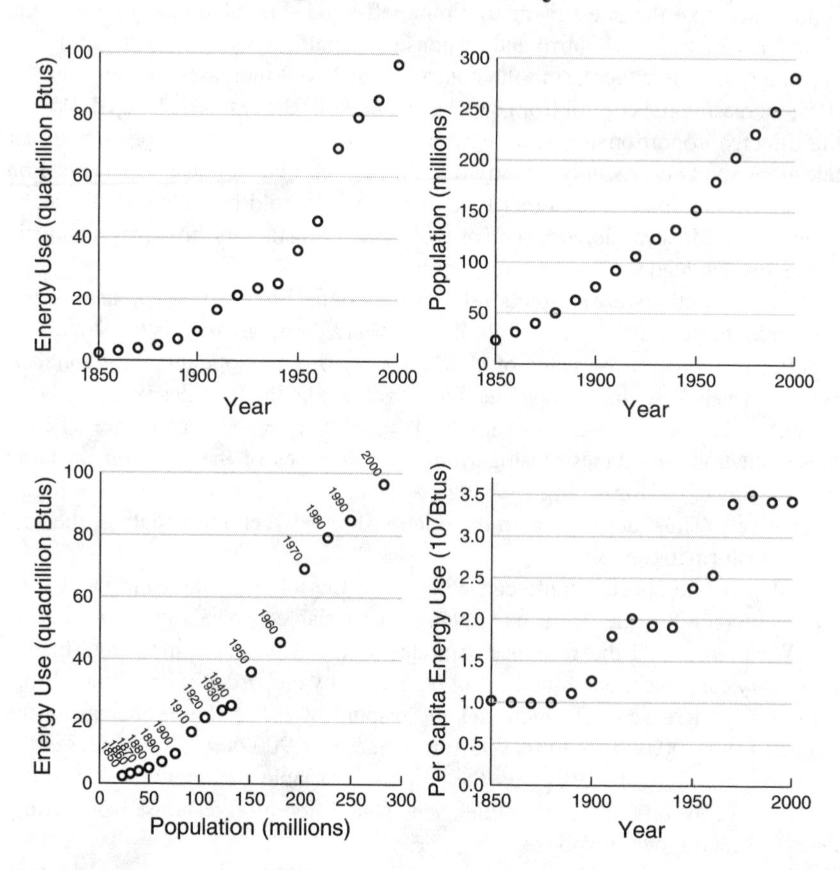

Figure A.5: In the top two graphs, I plot U.S. energy use and population from 1850 to 2000. Plotting the two dependent quantities against one another shows that they increase together with time, and plotting the per capita energy use shows a rather marked increase over the last century and a half.

Beyond any doubt, time is the ultimate independent variable. Here in Figure A.5 I plot two quantities changing with time, the human population and total energy use of the United States from 1850 to 2000. I discuss human population in detail in Figure 1.1 and U.S. energy use in Figure 3.1, and so don't dwell on them here. Both of the top graphs mesh well with the idealized dependent variable on the Y-axis and independent variable on the X-axis, but those graphs provide limited information.

Here I want to make two points, one scientific, the other a minor pedagogical one. At bottom left I eliminated the independent variable, time, and plotted the two dependent variables, energy use and population, against one another. Pedagogically, scientists don't always know whether the variable plotted on the horizontal axis really has the status of an independent variable, nor whether the one on the vertical axis really qualifies as a dependent one. Sometimes, as I've done here, people plot co-dependent variables against each other because that's all they have to plot. Indeed, Figure A.3 shows an imaginary situation where one variable predicts just 8% of the variation in the other variable. Take that small bit away and there's no correlation whatsoever between the variables, meaning we have two essentially independent variables. In other words, fixating too intently on the idea of an independent X-axis variable doesn't carry you very far.

That said, my more concrete point is that people use energy; thus I chose population as the new "independent" variable on the X-axis that drives energy use. My curiosity about per capita energy use through time causes me to divide total energy use by population, revealing that Americans doubled their energy use between 1850 and 1950, and have nearly doubled it again since 1950.

Graphs allow you to manipulate and explore data in ways that just aren't possible by staring at a table of numbers. Graphs allow you to pose questions and answer them with the available data, or contemplate the additional data needed to address new questions. In other words, they let you think like a scientist.

Sample sizes and measures of variation.

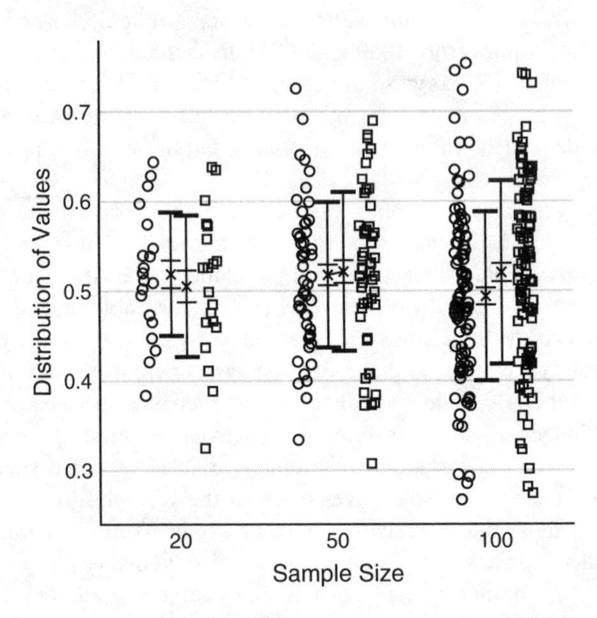

Figure A.6: A fundamental difference exists between a distribution's standard deviation and the standard error in a sample of points. Circles and squares arise from two distributions having the same standard deviation, 0.1, but different means, 0.49 and 0.51, respectively. Sampling just 20 points from each distribution results in rather poor estimates of the standard deviation, given by the thick bars, and overlapping standard errors, the thin bars, that do not resolve the differences between the distributions' means. Increasing the sample sizes to 50 and 100 provides better estimates of the standard deviations, and the standard errors shrink enough to distinguish the differing means.

Imagine some process that has an outcome governed partly by chance and partly by mechanisms — something more complicated than just a coin flip with a heads or tails outcome, more like flipping 100 coins, which, on average would produce 50 heads and 50 tails. A distribution of outcomes like that one has a mean, or average value, and some width, like a bell curve. A standard deviation — which mustn't be confused with a study's standard error (explained below) — provides the most common measure of the distribution's width. If a scientist undertakes a single measurement from an experiment whose outcome follows a particular distribution, about 68% of the time the experiment's result will fall between the mean minus the standard deviation and the mean plus the standard deviation.

I produced Figure A.6 using a computer program to sample two different "normal" distributions,[5] and again think bell curves. My distributions, which I've plotted as circles and squares, have slightly different means, 0.49 and 0.51, respectively, and identical standard deviations, 0.1.[6]

The standard deviation of a distribution is very different from the standard error in a sample of measurements. The former involves the width of the distribution, and the latter estimates how accurately an experiment estimates the distribution's mean. In the figure I show, between (and alongside) the circles and squares for each set, the averages, standard errors, and standard deviations. Each "x" marks the sample average, the thinner error bars closest to the x delimit the standard error, and the thicker error bars at the ends mark the standard deviation.

First, notice how the standard deviation error bars change length between sample sizes 20 and 50, but hardly at all between sample sizes 50 and 100. Experiments with large sample sizes provide good estimates of the underlying distributions, at least as measured by distribution width.

Second, standard errors decrease as an experiment's sample size increases. Statisticians proved that an experiment's standard error equals the underlying distribution's standard deviation divided by $\sqrt{N}$, where N equals the sample size. Given this connection, quadrupling the sample size only halves the standard error.

Third, the standard error represents how accurately an experiment estimates the distribution's mean. For example, compare the means and standard errors from the $N = 20$ experiment with those of the $N = 100$ experiment. Small sample size experiments don't estimate means terribly well, but their standard errors cover the actual means, well estimated by the large sample size experiment.

Finally, scientists invoke statistical tests to decide whether two averages differ; for example, are there different social benefits in barren versus green environments? These tests try to distinguish differences between the estimated averages for the $N = 20$ experiment, where detection is impossible, and the $N = 100$ experiment, where the difference is clear.

Notes

Chapter 1: Cities and Nature

(Figure 1.1, p. 4) Human population increased sixfold over the last century.

[1]The U.S. Census Bureau summarizes historical population numbers from several sources at www.census.gov/ipc/www/worldhis.html.

[2]I make many conversions from one unit to another in this book, for example, from people per square kilometer to people per square mile, and here's how to do it. We all know that 1 kilometer equals 0.62 miles. (Actually, few people have all these numbers stored in their head, including me, and I find them somewhere online, for example *onlineconversion.com.*) That equivalence means the fraction (1 kilometer)/(0.62 miles) equals one. We can multiply any number by one without changing its value, so write our global population density as (40 people/km^2)(1 km/0.62 mi)^2. The km^2 in the numerator cancels with that in the denominator, and $40/(0.62)^2=104$. Thus, our global population density equals 104 people per square mile.

(Figure 1.2, p. 6) Water, warmth, and light make plants grow.

[3]Besides capitalizing on the difference between potential and actual evapotranspiration in deserts, farmers gain at least one further benefit when irrigating desert fields — nearby noncrop areas support few bugs and weeds that recruit to crop plants. Desert farmers' crops need fewer pesticide and herbicide treatments.

[4]Many people have trouble reading graphs; scientists who teach nonscientists know that, but because graphs transfer so much information so efficiently, graphs just won't go away. I've added an appendix to this book covering basic graph-reading skills. A number of websites run by scientific associations provide guidance into these troubles. For example, concerns for teachers are at tiee.ecoed.net/teach/essays/students_interpreting_graphs.html, and from more of a student's perspective, see tiee.ecoed.net/teach/essays/figs_tables.html. Here's what you do: Break reading the graph into two steps, describe and interpret. Break describing into four parts: (1) Examine the axes and understand the variables; (2) look at the units used to measure the variables; (3) determine the symbols or lines used for the variables; and (4) see the patterns made by the variables. Some questions you ask yourself while looking at patterns: Is it steady or fluctuating? Are fluctuations random or systematic? Do the data trend upward or downward, and is the trend stronger than the fluctuations? No important scientific discoveries are really accepted until other independent

scientists double-check the information provided in these graphical descriptions. In the next step of graph-reading, you interpret what the description tells you, and you're allowed to be skeptical and bring in other information that the graph-maker might not have known or hasn't told you.

[5]Cleveland et al. (1999) provide the data showing how net primary productivity increases with evapotranspiration. It's unclear whether their numbers refer to potential or actual evapotranspiration.

[6]The plant growth versus evapotranspiration data in 1.2 has $N = 23$ points, and the fit has a correlation coefficient of 0.81 to the fitted function $y = 2.2121 + 5.7865x$.

[7]Potter et al. (2006) look at the United States' carbon budgets, including net primary productivity (NPP). Hicke et al. (2002) examine trends and average North American NPP.

[8]Typically, the term *species* refers to a genetically isolated, interbreeding set of organisms, but the concept becomes stressed because some sets are isolated only by space, others by behavior, and still others can reproduce offspring with much reduced fitness.

[9]Tree species richness increasing with evapotranspiration plot comes from Currie (1991). My fit for the tree species vs. evapotranspiration graph in 1.2 has $N = 258$ points, a correlation coefficient of 0.813, and a fitted function of $y = -13.776 + 0.1494x$.

(Figure 1.3, p. 8) High evapotranspiration promotes biodiversity.

[10]The plot showing species richness increasing with evapotranspiration comes from Currie (1991).

(Figure 1.4, p. 10) Humans exceed the natural population density for their body size.

[11]R.H. Peters wrote one of the best and oldest books, published in 1983, on the topic of body size scaling. I've shown the graph for carnivorous animals, but a very similar graph exists for herbivorous animals. I'm not aware of a fully accepted mechanism behind the correlation seen between population density and body size.

[12]Auerbach and Ruff (2004) use over 1,000 fossil skeletons, estimating prehistoric human body size to be about 50–60 kg, which weighs 110–132 pounds.

[13]Pozzi and Small (2001, 2005) describe the details behind the distribution plot of block-level 1990 U.S. populations. They categorized populations as rural (less than 100 people/km^2), urban (more than 10,000 people/km^2), and suburban in the middle range. I learned of this plot in a 2009 book by the National Research Council, *Urban Stormwater Management in the United States*.

[14]What about the population sizes for some of the species we have domesticated? The cattle population in the United States over the last few years, 2004 to 2007, has been about 100 million (USDA National Agricultural Statistics Services). There were about 453 million chickens in December 2006, worth about $2.60 each. With reference to these domesticated animals' population numbers, their life spans, of course, are unnaturally short, meaning that these population numbers are underestimated for their body sizes.

[15]Here's a different way to look at prehistoric densities. If our world population of 6.7 billion people was a natural density, then we should be a 1-lb (1/2 kg) carnivore (perhaps a large weasel), or a 6-lb (3 kg) herbivore (an average groundhog).

(Figure 1.5, p. 12) The last century brought increased agricultural efficiency.

[16]Crop yield and price data comes from www.nass.usda.gov/QuickStats.

[17]When I use the phrase "biological engineering," I'm also thinking about old-fashioned plant breeding. Certainly we recently developed much better ways of putting desired traits into many

types of organisms, but people have been doing that for a long time.

[18]MacDonald et al. (2006) provide the USDA farm breakeven numbers.

[19]A U.S. bushel is an 8-inch-deep cylinder of diameter 18.5 inches, a volume of just over 35 liters, and is equal to 4 pecks. It is now defined as a specific mass or weight of the commodity in question; for shelled corn that weight is 56 pounds at 15.5% moisture content.

[20]An acre equals 4,840 square yards, 43,560 square feet (a square about 209 feet on each side), or about 4,046 square meters. One square mile contains 640 acres, equal to a section. An acre also equals a chain (22 yards) by a furlong (220 yards). An acre just outsizes a football field, and a city block covers about 4 acres. Finally, 1 hectare is a square 100 meters on a side and equals 2.47 acres, and 100 hectares equals one square kilometer.

[21]This idea of 10% annual returns was written before fall 2008, before the collapse of the stock market.

(Figure 1.6, p. 14) Fossil-fuel-based nitrogen production increased crop yields.

[22]Herbivory and/or plant death happens, and decomposition leads to ammonium ions, NH_4^+, and ammonia, NH_3. Nitrification, called the oxidation of ammonium, transforms these chemicals to the nitrate ion, NO_3^-, and many bacteria make their living using the energy released in this transformation. Plants take in nitrate ions (and some ammonium, too), then put the energy back into these molecules to make ammonium ions. This process is called reduction. Denitrification transforms nitrate, NO_3^-, into nitrogen gas, N_2, and nitrous oxide, N_2O, releasing nitrogen to the atmosphere.

[23]Citations to ancient Greece and Roman agricultural practices are found in White (1970).

[24]Schmer et al. (2008) report that *real* farms growing switchgrass produce five times more energy than consumed, making switchgrass a potentially efficient energy crop.

[25]Presterl et al. (2002) studied nitrogen limitation in corn hybrids, comparing low and high nitrogen hybrids at low and high nitrogen levels.

[26]Considering switchgrass, note that in Figure 1.6 10 Mg/hectare biomass equals 1 kg/m^2, right around the values shown for plant growth in Figure 1.2 (except carbon versus dry biomass — multiply by 0.44 to convert to carbon [Dan Walters, pers. comm.] makes it spot on), and just over the grain produced by corn (ignoring stalks and such). Switchgrass isn't a miracle plant, but grows without much cultivation and might not take land for food production out of commission.

[27]Vogel et al. (2002) studied switchgrass biomass production with increasing nitrogen.

[28]The high nitrogen treatment added enough fertilizer to get to a level of 200 kg/hectare (179 lbs/acre) of available nitrogen. This number is agriculturally relevant: Corn growers apply nitrogen fertilizers at about 140 lbs per acre in Minnesota.

[29]A plowed-under crop of alfalfa can add up to 300 kgs of nitrogen per hectare, but, of course, at the cost of one year's crop. Crop rotation with soybeans helps soil quality while also producing a marketable crop. Nitrogen can fall from the sky when nitrogen gas gets converted to a useful form, but the amount is minimal. Rain brings in about 7 kgs of available nitrogen per hectare per year, or about 6 lbs per acre. In contrast, the air above an acre holds about 35,000 tons of nitrogen gas, N_2, essentially unavailable to all but a few biological processes.

[30]Galloway et al. (2003) document the growth of nitrogen use over the last century.

(Figure 1.7, p. 16) Urban land use grew as small farms disappeared.

[31]Many categories of land exist that complicate the land-use change picture. For example, the total for wilderness-based "special use" increased from roughly 61.4 million acres in 1959 to

242.2 million acres in 2002, but much of that land (58%) is located in Alaska. All told, here's another way to look at human population density. There are about 2.3 billion acres of land in the United States and a population of about 300 million, or about 8.7 acres per person. The 48 states (excluding Alaska and Hawaii) have 1.9 billion acres and a population of about 298 million, or about 6.4 acres per person. In contrast, the world's population of 6.7 billion on Earth's terrestrial surface of about 38 billion acres averages to about 5.8 acres per person across the globe.

[32]Lubowski et al. (2006) provide the land-use data through a U.S. Department of Agriculture bulletin. This bulletin shows that there are many technical issues concerning classifying land use.

[33]Two publications from the U.S. Department of Agriculture, MacDonald et al. (2006) and Hoppe et al. (2007), provide great amounts of information on U.S. farms, including the farm size distribution and economic viability data I show here.

[34]I have not examined or considered the amount of foreign land that Americans rely on for agricultural production, but certainly importing food plays an important role in U.S. land-use change.

(Figure 1.8, p. 18) Cities change ecological communities.

[35]Bock et al. (2008) studied bird species richness in southeastern Arizona.

[36]Hansen et al. (2005) compiled the cross-taxa surveys of species richness versus land-use types. Richness values are measured relative to the highest observed species richness across all of the listed land-use types.

[37]Donnelly and Marzluff (2006) performed an extensive study of bird species richness in Seattle, Washington.

[38]Ostfeld and Keesing (2000) provide a wonderful overview of the ecology surrounding Lyme disease and cite 10,000–17,000 cases reported in the United States in the 1990s. Allan et al. (2003) provide the data regarding infected ticks with forest patchiness.

[39]These ticks go by a common name, deer ticks, because the adult tick often takes its final blood meal from a white-tailed deer, but by no means does it only parasitize deer.

(Figure 1.9, p. 20) Agricultural and urban land use reduce streamwater quality.

[40]Salt concentrations in streams are discussed by Kaushal et al. (2005). Data for Figure 1.9 were kindly provided by Sujay Kaushal and the Baltimore City Department of Public Works and Hubbard Brook Ecosystem Study groups. The reservoir data involved the Liberty Reservoir, specifically sampling station MDE0026.

[41]Most cities treat stormwater differently than sewage, meaning, for example, that Durham doesn't treat stormwater at all. (Most municipalities separate these water issues, but not all, and one exception is New York City.) No one can reasonably expect water treatment plants to handle a couple inches of rain over a few square kilometers of impervious surfaces during a two-hour rainfall. The water volumes are immense. However, the first tenth or quarter inch of runoff (called the first flush) has higher pollution levels and warrants special treatment.

[42]Kaushal et al. (2008) provide the nitrogen data around Baltimore, Maryland, for concentrations versus impervious surface fractions and exports from different land-use types.

[43]One can certainly question (as did a reviewer) whether farmers apply more nitrogen than necessary; if so, overapplication leads to more runoff than necessary.

[44]Peter Groffman made available the weekly sampling data allowing me to replot the Groffman et al. (2004) nitrate concentration plot.

(Figure 1.10, p. 22) Impervious surfaces in urbanized watersheds hurt organisms.

[45]Put a teaspoon of salt in a quart of water, and give it a taste; it's about one-quarter the saltiness of seawater. Now pretend you're a sensitive insect nymph living in it.

[46]Consequences of high salt concentrations from road applications are discussed by Kaushal et al. (2005).

[47]A large argument has ensued in the Durham–Raleigh area about which municipality should pay for the drinking water supply cleanup — the polluter or the user.

[48]Walsh and coworkers (2005a,b) provide in-depth coverage of urban stream health.

[49]Riley et al. (2005) demonstrated the rapid loss of sensitive stream species with urbanization, a natural measure of water quality. Sensitive species counted included species from the orders Ephemeroptera (mayflies), Plecoptera (stoneflies), and Trichoptera (caddisflies) that are intolerant to disturbed conditions. These insects have either a nymphal stage (with gradual molting into adult form) or a larval stage (followed by a distinct pre-adult pupal stage) that lives and feeds under water.

(Figure 1.11, p. 24) Durham rainfall exceeds regulated basin sizes.

[50]I found the 1997–2008 rainfall data from the National Climatic Data Center (ncdc.noaa.gov) for Durham's NWS COOP Station No. 312515. Data for this station go back to 1899, but only this more recent data was freely available. Note that some dates for the Durham rainfall data lack entries and I assumed zero rainfall as a conservative estimate. Data like these exist for stations located nationwide.

[51]I wrote a simple C computer program using the Durham rainfall data as input, determined the frequencies of rainfall amounts over these different periods, and then output the appropriate integrals over the probability distributions. The program is freely available from my website.

[52]The National Research Council produced an excellent overview of urban stormwater management, including a discussion of mitigation approaches (NRC 2008).

(Figure 1.12, p. 26) Reservoirs reduce sediments while providing water.

[53]Keep in mind the differences between the many types of "used" water: stormwater, wastewater or sewage, groundwater, drinking water, and reservoirs. Rain falls on impervious surfaces draining to streams, and it becomes stormwater that refills reservoirs. Rain falling on permeable surfaces soak in and can recharge groundwater sources, though some rain runs off permeable surfaces, too, and can make its way to streams. Reservoirs recharge from water flowing through streams, which can mean stormwater flushed into urban streams. Wastewater (or sewage) is the stuff flushed down toilets, showers, and sinks, partly cleaned in treatment plants before entering streams. Yet another term is graywater, water coming from sinks, washing machines, and, in some jurisdictions, also showers and bathtubs. Groundwater comes from wells drilled into aquifers, which get recharged by rainwater filtering through the ground throughout watersheds. Drinking water can come from both reservoirs and groundwater sources.

[54]Cooper et al. (2004) present the work on sedimentation rates in Pamlico Sound. They also show changes in the ecological community through several centuries as the stormwater sedimentation changed.

[55]Water use for each county was found at water.usgs.gov/watuse/data/2000/index.html. Another interesting calculation is how much of our total rainfall we humans use, specifically Americans. NOAA reports that the total precipitation for 48 states averaged about 29 inches per year over the last 100 years (www.ncdc.noaa.gov/img/climate/research/2006/ann/us-summary.html). The United States Geological Survey reports that water use across United States equals about 408 billion gallons per day for all uses from drinking

water to power plant cooling (pubs.usgs.gov/circ/2004/circ1268). Sometimes these uses reuse water already used for another purpose, but this latter number means 1.5 x 10^{14} gallons per year, or 5.5 billion acre-inches. The 48 states have about 1.9 billion acres, meaning we use about 2.9 inches of water covering that entire area each year. That's our water footprint. In other words, humans use 10% of the U.S. rainfall, a bit more than Durhamites use.

[56]This water-use calculation can go in many directions. Here's another. Divide the 408 billion gallons per day number by the U.S. population, 300 million, to get 1,360 gallons per day per person. That's six times more than I calculated in the text for Durham citizens, leading to about 18 inches of rain per year, or nearly 40% of Durham County's rainfall. That's a seriously big fraction. The difference represents personal versus nonpersonal use.

[57]Yet another interesting feature of Durham County: Its urban streams help fill Falls Lake, a reservoir serving drinking water to Wake County, home to Raleigh, North Carolina. Preservation of open space and retrofitting impervious surfaces for better stormwater infiltration in Durham County becomes an important issue for another county's citizens. Who pays? Although the Army Corps of Engineers acquired much of the land immediately surrounding the reservoir, these results imply that further development of upland areas reduces water quality.

[58]Weiss et al. (2007) compare many types of stormwater treatment approaches, concluding that constructed wetlands are the best solution. These types include "dry extended detention basins" (a barren hole) that hold water for less than two days; "wet retention basins" (a pond) that might hold water almost continuously; "constructed wetlands" (pond and meadows) that hold water for an intermediate time; and "infiltration trenches" (parking lot strips), which are pretty much anything that filters stormwater and prevents it from entering streams. The tree-planting locations in the Food Lion parking lot (see Figure 2.12) could act as an infiltration trench if only the surrounding curbs were removed to allow runoff to reach those permeable areas.

Chapter 2: Shading and Cooling in City Climates

(Figure 2.1, p. 30) Low vegetation correlates with high temperature in Durham.

[1]When rain hits an impervious surface, it doesn't directly seep into the ground. That elsewhere might be the side of the road, alongside a building, into rain barrels, or down a stormwater pipe.

[2]An excellent overview of the urban heat island is given by Pickett et al. (2001). The first published mention of urban heat islands, at least that I'm aware of, comes from temperatures measured in and around London by Luke Howard, in an 1833 book titled, *The Climate of London, Deduced from Meteorological Observations, Made in the Metropolis, and at Various Places around It.* These temperatures, taken from 1807 to 1816, show an urban temperature of a degree or two Farhenheit higher than the surrounding countryside.

[3]Canopy and temperature maps of Durham, North Carolina, were provided by Joe Sexton using the magic of geographical information system (GIS) software. The canopy map uses 30 m square pixels, and the temperature map uses 120 m square pixels. Color versions of these images are available and provide more detailed representations. Be aware that urban heat islands don't happen every day. Thermal images from fall, winter, and early spring show absolutely no sign of the city, and even a few days of wet, overcast summer weather leaves no evidence of an urban heat island.

[4]An extensive review on the urban heat island by Arnfield (2003, p. 7) cites numbers ranging from 11 to 1,590 W/m^2 for human energy input. The highest level was for Tokyo in the wintertime. My rough, county-level-averaged number of 83 Btu/m^2/day=1W/m^2 is an order of magnitude

lower than this range's lower bound. My energy calculation makes the fundamental assumption that people are spread out uniformly, but in city cores maybe this assumption doesn't hold true. The City of Durham has three times the density of the county, and the city core is even denser; that probably brings my number up to the lower levels of this range. Even so, human inputs in Durham are small, maybe 5%, compared to solar radiation input of 13,600 Btu/m^2/day, or around 160 W/m^2. Perhaps having 100 times more people than average in a tall office building results in energy use heat inputs 100 times greater, which multiplying out gives 8,000 Btu/m^2/day, approaching that of solar input. Interesting plots might be energy use versus temperature, and population density versus temperature, though correlations between energy use and industrial/ commercial land use would likely get in the way.

(Figure 2.2, p. 32) Low vegetation correlates with high temperature in Indianapolis.

[5]Vegetation and temperature correlation images from Indianapolis, Indiana, were provided by Jeff Wilson, and are discussed in depth in Wilson et al. (2003). Another set of images showing temperature and land use for Indianapolis, can be found in Weng et al. (2004), and shows the same correlation as vegetation. It seems likely that places with low vegetation in a city are simply the places where there are high industrial and commercial densities because you can't have two different things in the same place.

[6]Vose et al. (2003) describe and review the normalized difference vegetation index (NDVI) calculation very well.

[7]Arnfield (2003) reviews urban boundary layer science.

[8]Yet another example of an urban heat island comes from Hydarabad, India, a city of nearly 4 million people, where satellite measurements indicate a UHI of about 7C (Badarinath et al. 2005).

(Figure 2.3, p. 34) Low temperature correlates with high vegetation.

[9]Use of satellite-derived information for guiding subdivision impacts is discussed in Wilson et al. (2003).

(Figure 2.4, p. 36) Urban heat islands spawn thunderstorms.

[10]Thunderstorms produced by the UHI over Atlanta, Georgia, are discussed by Bornstein and Lin (2000).

[11]Given an air temperature with some humidity level, the dewpoint is the temperature that the air needs to be cooled to for the water to condense out of the air, or in other words, have a relative humidity of 100%.

[12]Dixon and Mote (2003) discuss thunderstorm generation associated with UHIs in Atlanta. Grady Dixon kindly provided the data to produce Figure 2.4. He describes UHI-induced thunderstorm conditions: "When the air is more humid, the UHI is usually less pronounced because the maximum temperature for the day is not as high. However, I found that it is those days when UHI-induced thunderstorms are most likely to occur. The plot showed that greater low-level moisture, in particular, was important for distinguishing between days with or without UHI-induced thunderstorms. However, a cool, humid day will not yield UHI-induced thunderstorms while a warm, humid day likely will."

(Figure 2.5, p. 38) Cities change rainfall patterns.

[13]Oke (1982) describes the time course of the urban heat island. Urban areas experience lower radiative heat loss in the early afternoon, but higher losses in the evening and overnight as im-

pervious surfaces slowly release heat. This process also describes how the typical thermal mass behaves in passive solar homes.

[14]Burian and Shepherd (2005) studied rainfall changes between "pre-urban" and 'post-urban" years in Houston, Texas. The data shown here covered June, July, and August, over the time span 1984–1999.

[15]Information on the rainfall-measuring satellite can be found at trmm.gsfc.nasa.gov.

[16]One of the earliest mentions of weekly patterned rainfalls comes from Ashworth (1929, 1944) involving measurements from London between 1898 and 1943. In that situation, coal-burning factories shut down on Sundays, the day with the lowest precipitation level over the measured period.

[17]Bell et al. (2008) present the weekly morning–afternoon rainfall pattern across the southeastern United States.

(Figure 2.6, p. 40) Lightning strikes reflect urban weather changes.

[18]Tony Stallins provided Figure 2.6 showing the frequency of lightning strikes in Atlanta, Georgia. The work is discussed in detail in Stallins (2004), Stallins and Bentley (2006), and Stallins et al. (2006).

[19]The National Lightning Detection Network, a system that measures and locates every lightning strike, is described at thunder.nsstc.nasa.gov.

[20]I obtained the Durham County lightning image, with great appreciation, from Vaisala, Inc., the lightning detection equipment manufacturer and data services provider that owns and operates the U.S. National Lightning Detection Network. I do not have the data itself to test whether or not a correlation exists between lightning strikes and urbanization, but that would be the next step.

[21]Stallins (2004) cites the economic impact of lightning in Georgia and the United States.

(Figure 2.7, p. 42) Cities grow warmer.

[22]The historical temperature plot conceptually follows one by Akbari et al. (2001), but with more extensive Los Angeles historical temperatures provided by the National Oceanic and Atmospheric Administration at www.wrh.noaa.gov/lox/climate/cvc.php.

[23]All of the regressions provide highly significant fits.

[24]The early work on the urban heat island (UHI) and city size originated with Oke (1973), and was further discussed by Oke (1982). Souch and Grimmond (2006) recently published a detailed review of more recent work on UHIs. Here I discount the truly historical 1833 work by Luke Howard!

[25]Can the reason behind the UHI increasing with city size be as simple as a bigger chunk of concrete holds more heat? Some graduate student should find data on the total impervious surface as a function of population size and see if the UHI scales in a correlated way. There's a large amount of detailed work taking place that gets at the basic science involved; one review of very recent work is by Souch and Grimmond (2006).

(Figure 2.8, p. 44) Closed-in urban areas have higher heat islands.

[26]The fusion reactions taking place within the Sun are much, much hotter, but that energy gets absorbed by the mass of the Sun itself and is then reemitted as light at cooler temperatures. On Earth we see and feel the radiation coming off the Sun's surface at an almost cold-and-frozen, relatively speaking, 6000C. Scientists often refer to light as being at a certain temperature, though that's a bit informal. What's meant is that the atomic and subatomic particles of objects at a certain temperature radiate photons (i.e., light), and the energy distribution of those photons depends on

the object's temperature.

[27]It is true that all frequencies of light have the same speed in an absolute vacuum but through various media that truth breaks down. Think about light going through a prism, and how white light splits into a rainbow. That demonstration shows a frequency-dependent speed of light in the medium making up the prism.

(Figure 2.9, p. 46) The urban heat island may be weak while Earth warms.

[28]Peterson (2003) argues that the UHI is due mostly to measurement biases, but recently concedes very small UHI temperature increases (Peterson and Owen 2005).

[29]Tom Peterson suspects, but hasn't shown yet, that the difference between thermal satellite images and weather stations is that weather stations are more likely to be placed in parks than in parking lots, meaning the difference between stations and satellites is revealing underlying variation in temperatures.

[30]UHIs were examined between the west and east coasts of the United States by Sun et al. (2006).

[31]Parker (2006) provides a nice summary of the literature pursuing this urban heating possibility for the global warming trends, considering calm and windy days as a test. There appears to be very little evidence to support the UHI-induced alternative idea. If true, calm days should have large UHIs as urban air temperatures rise, but on windy days, just as with a car's radiator, cool rural air should blow away warmed city air, leaving small UHIs. Looking at data for those days separately demonstrated increasing trends in both cases, on a global scale, similar to the pictured L.A. trend. The results falsify the idea that the UHI plays a major role in the measurements of global warming because data taken in the absence of an UHI still shows the global trend.

[32]For an older study of paired urban and rural stations that concluded urban heat islands exist, see Karl et al. (1988).

(Figure 2.10, p. 48) Equal heat contained in air, a sprinkling of water, and an asphalt road.

[33]The densities of asphalt and concrete are about 2,200 and 2,400 kg/m^3, respectively, and for simplicity I'll assume 2,000 kg/m^3 for both. For comparison, dry soil has a density of about 1,000 kg/m^3. Asphalt, clay, and concrete specific heat values are about 0.19–0.22 cal/g/C, which means it takes about 0.2 calories of heat to increase the temperature of 1 gram of these materials by one degree Celsius. One British thermal unit (Btu) equals 252 calories and also equals 0.293 watt-hours. (One Btu is the amount of heat required to warm up one pound of water one degree Farhenheit at 60F.) For comparisons, burning a 100 Watt incandescent bulb for an hour uses 100 watt-hours of energy, about 341 Btu, or about 86,000 calories. That many calories would heat up a liter of water — 1,000 grams, about a quart — by 86C, or from 14C, cool water, to 100C, which is the boiling point of water. In other words, burning a 100W incandescent bulb for an hour, or heating up water for pasta involves the same energy demand.

[34]I leave multiplying out the numbers in the table as an exercise for the reader! Another heat content calculation like the ones in Figure 2.10 shows that 2,400 Btu would heat up a 1-inch (2.54 cm) rainfall — about 25 liters of water spread over one square meter — from 20C up to a very warm temperature of 44C. In reality, of course, both the road and the water would equilibrate to some common temperature. That's a lot of water needed to cool things down, but trees are cool because their transpiration makes a very important contribution to cooling, just as does the sprinkling of vaporized rain.

(Figure 2.11, p. 50) Whiter surfaces are cooler.

[35]Pavement and material temperatures against albedo were reported by Pomerantz et al. (1999).

[36]Loss of the oily tar portion of pavement is the cause of the pavement bleaching with age, which heat and sunshine accelerate. Loss of the tar means the small stones are bound together less tightly and the road deteriorates (McPherson et al. 1999). It was estimated that in San Joaquin Valley communities in California in 1999 the cost of resealing pavement was about $3 per square meter, unshaded pavement needed resealing every 10 years, and shaded pavement every 20 or so years.

[37]Temperature profiles in the Saudi Arabian desert were provided in Al-Abdul Wahhab and Balghunaim (1994). In their study they drilled out cylindrical cores from roads, implanted temperature sensors at various depths, and then reinserted the cores. Their motivation arose from preventing vehicles making ruts in hot, desert roads, not urban heat islands. These numbers agree roughly with model predictions for temperature profiles derived in Nobel (1991). That model provides two interesting predictions. First, the $1/e$ damping depth for daily temperature variations is 9.1 cm (and 1.7 m for annual temperature variations). Second, given a 15C amplitude in daily temperature variation (30C difference between maximum and minimum temperatures), a 1C amplitude variation takes place at a depth of 25 cm. Thus, I argue that a 15 cm depth is a fine approximation for uniform heat storage.

(Figure 2.12, p. 52) Parking lot trees could provide shade.

[38]McPherson (2001) examined parking lot shading by trees in Sacramento, California, and presents ideas for redesigning parking lots for enhanced shading.

(Figure 2.13, p. 54) Big trees could provide lots of shade.

[39]Tree data for crown projection area is given by Shimano (1997), who cites an inaccessible paper by Tatewaki et al. from 1966 concerning temperate forests in central Japan.

[40]These two crown size graphs seem to disagree, but the lack of competition in the right plot explains the difference. For example, look at a tree of 75 DBH. The Shimano (1997) data suggest a crown projection of, say, 100 m^2, but the other studies suggest a diameter of around 15 m. Since the area is $\pi/4D^2$, the latter crown projection area ought to be about 175 m^2. Those numbers seem quite different, meaning trees without competition spread out a lot.

[41]Hasenauer (1997) looked at tree shape in Austria, and Smith et al. (1992) looked at pine species relevant to the southeastern United States.

[42]Two of these trees were cut down shortly after I took the measurements; I hope no cause and effect was in play.

[43]The data point plotting the Food Lion parking lot trees have a dot inside the circle; that dot is a different data point, not the parking lot trees.

(Figure 2.14, p. 56) Bigger and younger trees transpire more water.

[44]Vose et al. (2003) and Wullschleger et al. (1998) provide sap flow data for different places and tree species.

[45]Vose et al. (2003) present transpiration data for trees as functions of light and heat, and Lu et al. (2003) show evapotranspiration versus precipitation in southeastern U.S. watersheds.

[46]An even broader scale measure comes from Hoover (1944), who reports measurements before and after a logging operation in the North Carolina Appalachian Mountain range. Results show that the annual transpiration by the intact forest totaled 48.7 cm over an entire year, and dividing

by 365 days gives an average of about 1.3 mm per day.

(Figure 2.15, p. 58) Trees near asphalt stop transpiring early in the day.

[47]Kjelgren and Montague (1998) measured tree temperatures, stomatal conductance, and water loss over paved and turf surfaces, and also considered ash and maple trees.

[48]Georgi and Zafiriadis (2006) performed the study of shade trees in Thessaloniki, Greece.

(Figure 2.16, p. 60) Evapotranspiration is high from watersheds and lawns.

[49]The fit for the evapotranspiration versus rainfall graph in Figure 2.14 has an R^2=0.32 with the fit $y = 27.3 + 0.431x$.

[50]Ebdon et al. (1999) performed a study of evapotranspiration in lawns.

[51]Transpiration of 1 kg/day/m^2 equals 1 mm/day, so 4 kg/day equals 4 mm/day. This equivalence is best understood by noting that one liter of water $(10 \text{ cm})^3$ has a mass of 1 kg. Imagine a square box 10 cm on a side, cut it into 100 slices, each 1 mm thick (10 cm = 100 mm). Those 100 slices cover one square meter. The evapotranspiration value quoted here agrees well with the averages reported by Carrow (1995), pointed out by an anonymous reviewer, for several grass species in the southeastern United States, 3.03–3.80 mm/day, with some experiments exceeding 6 mm/day.

[52]Heilman and Gesch (1991) examined the temperature reduction turfgrass provides to nearby buildings.

[53]McPherson et al. (1989) demonstrated that rocky landscaping was the most cost-effective option in Tuscon, Arizona.

[54]Givoni (1991) reviewed many studies that use vegetation to provide cooling benefits.

(Figure 2.17, p. 62) New developments can plan for shade.

[55]Mennis (2006) discusses subdivision age as a factor predicting vegetation.

(Figure 2.18, p. 64) Paving and grass can be combined.

[56]Brattebo and Booth (2003) studied permeable paving's water retention and pollutant filtering abilities over several years, finding extremely favorable results.

[57]Walsh et al. (2005a) describe negative consequences on urban streams from paving.

Chapter 3: Energy Use and Carbon Budgets

(Figure 3.1, p. 68) U.S. energy sources have changed.

[1]Shafik and Bandyopadhya (1992) discuss income and environmental quality. Measured over global scales, increasing per capita income generally shows an initial deterioration in many environmental quality indicators, but with further increasing income, most indicators show increasing environmental quality. Perhaps this relationship reflects a use of wood as a primary energy source in developing countries, and transitioning out of biomass fuels as they develop.

[2]Before 1950, the human emissions signature was observable in a diluted C_{14} to C_{12} ratio, but atmospheric nuclear weapons testing obliterated that signal by putting tremendous amounts of radioactive carbon into the atmosphere. These topics are covered by Druffel et al. (2001) and Levin and Hesshaimer (2000). This dilution is called the "Seuss effect."

[3]Energy source data come from the Annual Energy Review 2006, along with 2007 and 2008 numbers from the 2008 publication, produced by U.S. Department of Energy, Energy Information Administration DOE/EIA-0384. This publication provides a huge amount of fascinating information.

[4]Data on coal use and reserves are found on the U.S. Energy Information Administration website. They report both 1,000 million tons of coal per year and 23.79 x 10^{15} Btu of coal energy obtained: Simple division yields about 6,000 Btu/kg coal. (This number doesn't quite match up with numbers I present in Figure 3.16, which show 14,000 Btu/kg of wood, and with my statement that one unit of coal equals 1.59 units of wood. I have not yet resolved the discrepancy.) Their curves then represent our coal use over the last century as roughly 600 million tons of coal per year, or 60,000 million tons. U.S. coal reserves are estimated at 270,000 million tons — a few more centuries to burn, if we can stand the heat.

[5]Energy consumption statistics can be found at www.eia.doe.gov/oiaf/ieo/oil.html. Additional information can be found in the British Petroleum Statistical Review of World Energy 2008, at www.bp.com.

[6]To get some idea of our total energy use from domestic sources, let's just multiply rough averages for each of these annual energy use values by another rough estimate of how long each has been used. Looking at these curves, I estimate coal use as 15 quads for 100 years, petroleum use as 15 quads for 60 years, and natural gas use as 18 quads for 50 years. These numbers total 3,300 quadrillion Btus, and for further simplicity, let's say it was used evenly over the last 100 years, giving a rough value for the United States' energy use as 30 quadrillion Btus per year. Regarding my estimate of U.S. energy use as 30 quadrillion Btus per year, never mind that recent levels are more than 50 quads/year: In an order-of-magnitude calculation, 30 equals 50.

[7]Alaska's 1.7 million km^2 and 700,000 people really skew per capita calculations, for better or worse, with few people and lots of area (520 acres per person). Per capita perspectives of Canada, for example, would be very different from those of the 48 U.S. states.

[8]Average net primary productivity over the entire United States equals about 400 grams of carbon per square meter, and I use a conversion of about 14,000 Btus per kilogram of wood. Dukes (2003) provides some additional, independent connections between different fuel sources, including energy retention of NPP when fossil fuels form, and reports an energy content of plant material of 19,000 Btus/kg. Be warned: Other than this difference, I haven't checked my numbers against his results, either in this section or the others, so some discrepancies are likely.

(Figure 3.2, p. 70) States vary in their gasoline and electricity use.

[9]These numbers say that my family, a total of four people, consumes about 2,000 gallons each year, an amount that half-fills a 4-foot-deep, 15-foot diameter above-ground pool that holds 5,000 gallons. I'm sure my family doesn't directly purchase 1,000 gallons of gas each year, or two 10-gallon fill-ups per week, but we're certainly at half that amount. Add in schoolbuses, several gasoline-powered implements, and our air travel and this number becomes believable.

[10]Lifting 1 kilogram by 1 meter against gravity takes almost 9.8 Joules of energy (let's call it 10). Calories measure food energy, and more generally, things associated with heat, with 1 *C*alorie equal to 1,000 *c*alories, and one calorie can raise one gram of water by 1C (thus a Calorie raises 1 kilogram of water by 1C). (Note that the capitalization of the word *calorie* matters.) One Calorie equals about 4,200 Joules. Many home appliances measure energy use in Btus (British thermal units), which roughly equals 1,055 Joules.

[11]Dividing the land area of the United States, 10^{13} square meters (including Alaska), by the U.S. population, about 300 million, we have 33,000 square meters per person, or 8.2 acres. Excluding Alaska reduces the United States to 6.4 acres per person, or about 26,400 square meters per person.

[12]One gallon of gasoline equals about 120,000 Btus, about 43 Btus/gram, which works out to

60 million Btus for my 500 gallon consumption, or 1,800 Btus/m^2 over my 8 acres.

[13]The Annual Energy Review 2008 reports the rough breakdown of energy use, in quadrillion Btus, to be 21.6, 18.5, 31.2, and 27.9, for residential, commercial, industrial, and transportation, respectively.

[14]Kiehl and Trenberth (1997) and Maurellis and Tennyson (2003) provide wonderful discussions of solar energy. In other units, solar energy intensity at the top of the atmosphere is 342 W/m^2, and at the ground, 198 W/m^2. Conversions use 1 Btu of energy equals 1055 Joules, and the power unit, a Watt, equals 1 Joule/sec.

[15]Even if this solar panel area sounds reasonable enough, at least one 2008 cost-benefit study of solar panels demonstrated that they're not cost effective, even when factoring in nonmarket values (Borenstein 2008).

(Figure 3.3, p. 72) Per capita energy use depends on a state's population density.

[16]The fits for the two datasets in Figure 3.3, per capita gasoline, and electricity use versus state area, are both significant, and have, respectively, R^2 of 0.11 and 0.17; fitting the functional form, $y = A - B\ln(x)$, give As of 569.28 and 18,531, and Bs of -17.8 and $-1{,}209.2$.

[17]The idea that per capita energy use decreases with human density might not be borne out across all countries, however. During a talk in Bangalore, India, an audience member expressed surprise about this connection: In developing countries rural areas do not yet have electrical service (and perhaps have low fuel provisioning as well); thus they have little energy use. In contrast, the United States pushed rural electrification in the 1930s, a time when many European countries already had rural service. Thus, in that situation, per capita energy use increases with population density, and my results are conditioned on the assumption that everyone has access to the same energy resources.

[18]Unfortunately, the only information I could find about line loss was that electric companies allowed for a 7% loss due to transmission, but I found no information on variation across the country.

[19]State-level per capita electricity use can be found in the 2006 North Carolina Sustainable Energy Legislative Guide, prepared by the North Carolina Sustainable Energy Association, www.nc-sustainableenergy.org.

(Figure 3.4, p. 74) Economic productivity correlates with energy use.

[20]World Bank data were downloaded December 3, 2007, and differed drastically from data I downloaded on an earlier date. I do not understand the discrepancy, though the earlier data seemed to have grossly incorrect numbers. Energy use data plotted here were identified by their variable, **EG.USE.PCAP.KG.OE**, and I obtained the per capita GDP by dividing their GDP for each country, **NY.GDP.MKTP.CD**, by their population for each country, **SP.POP.TOTL**. A later visit to the site showed a seemingly new interface without these variables identified as such. My documentation may be outdated.

[21]"Purchasing power parity" is a more recent international comparison, describing economies from the ability of people to purchase goods. Much more information is available using GDP and GNP; thus I've used those measures throughout the book.

[22]The fit for the GDP–energy plot in Figure 3.4 is significant with $R^2 = 0.77$, fitting the functional form, $y = Ax^B$, and gives $A = 0.21$ and $B = 1.33$.

[23]Gross state product data available from the Bureau of Economic Analysis at www.bea.gov/newsreleases/regional/gdp_state/2005/gsp0605.htm.

[24]The correlation between a state's GSP and its population in Figure 3.4 seems too strong. I don't doubt that the state-level energy–economy connection likely arises from both quantities co-depending on population size and density, but the nearly perfect correlation makes me wonder whether GSP estimates come from combining state-level population sizes with per capita averages for various industries.

(Figure 3.5, p. 76) Photosynthesis links carbon, water, nitrogen, and sunlight.

[25]All amino acids contain nitrogen, and in long molecular chains, called polypeptides, make up proteins that fold up into chemically active components of biological processes. Parts of RNA (ribonucleic acid) and DNA (deoxyribonucleic acid) molecules are composed of nucleotide building blocks, each built out of a sugar molecule, a phosphorous-containing group of atoms, and a nitrogen-containing group of atoms.

[26]*The Biology of Plants* by Raven, Evert, and Eichhorn, discusses the elemental composition of plants. These few atoms comprise about 98% or more of nearly all organisms' weight.

[27]Carbon pools and fluxes are very nicely described in Chapter 11 of the biogeochemistry book by Schlesinger (1997).

(Figure 3.6, p. 78) Atmospheric CO_2 increased with human emissions.

[28]Six gigatons, the recent level of carbon emissions, equals 6×10^9 tons, or 12×10^{12} pounds, or 5×10^{12} kg. The atmosphere contains about 750 trillion kg of carbon, so emissions seem like a small fraction. However, Figure 3.6 shows that since 1950, the atmospheric concentration increased by about 20%, or about 150 trillion kg. Average annual emissions of around, say, 2×10^{12} kg for a period of 60 years totals about 120 trillion kg, a pretty close number.

[29]Of course, these historical CO_2 values have a coarse-grain measure over about 1,500 years, certainly longer time scales than we've been measuring over the last few decades. This results in something like comparing the EPA's 8-hour ozone levels versus 1-hour ozone levels. However, it seems tenuous to use that fact to comfort oneself that today's levels are just "natural" cycles, having nothing to do with the enormous amounts of fossil carbon we've released back from sequestration.

[30]Residence time of carbon atoms in the atmosphere is obtained by dividing the amount of carbon in the atmosphere, 750 trillion kg, by the amount absorbed by plants and oceans each year, roughly 200 trillion kg. That gives about four years. Another way to look at these numbers: If all the atmospheric carbon fell on the ground, there'd be about 1.5 kg/m^2, of which human contributions total about 0.4 kg/m^2, or about a pound per square yard. Yearly fossil fuel emissions into the atmosphere amount to 16 grams of carbon per square meter.

[31]Here's a rough calculation. If we could stop emitting all fossil fuels, the balance of all the carbon transfers would sum up such that around 3 trillion kg would be pulled out of the atmosphere each year. Dividing a rough estimate of the excess carbon in today's atmosphere, perhaps 175 trillion kg, by that sum gives 175/3 = 58 years. That's if all fossil carbon emissions stop.

[32]Some might argue that since humans have been around for the entire 400,000 years that the long-term graph covers, clearly we can survive varying CO_2 levels and temperatures. What's another few hundred thousand years, and if we already see such large natural variation, why worry? I believe those arguments are irrelevant to the current situation: Our present high population depends critically on growing food where we presently grow food. Rapid climate change could disrupt food production and imperil many people. Human-caused climate change, including global warming, is true, independent of whether one chooses to believe it to be so, but this topic falls

outside the purview of a book on urban environments.

[33]Land plants take in lots of CO_2 from the atmosphere, and the preponderance of the northern hemisphere's land mass makes it's seasonal plant growth observable.

[34]Glacial cycles in temperature and CO_2 result from something called Milankovitch cycles (Shackleton 2000). The Earth's orbit around the Sun isn't a simple circle; rather it has many ways to be different from a static, pure circle. I always fall back to thinking of a child on a swing. The swing is a pendulum with a "preferred" frequency. The child can pump the swing too fast, too slow, or just right to force the swing to swing. Now imagine a line of 10 swings of different lengths made out of different kinds of ropes and chains, and the swings are linked together one to the next by rubber cords. Ten children pump these swings and out pops an overall swinging of the swings. In a similar way, all of the variability in Earth's orbital motions, put all together, leads to climate cycles on Earth having 100,000-year cycles.

[35]Falkowski et al. (2000) present the analysis of CO_2 periods in historical records.

(Figure 3.7, p. 80) Global warming changes nature.

[36]Brohan et al. (2006) present the temperature change over the last century. I obtained the data for my plot from www.cru.uea.ac.uk/cru/data/temperature, which provides an excellent discussion of the data's origins. The data provided monthly deviations (from the 1961–1990 average) for every year, along with the yearly averages; I plotted the yearly average with error bars represented by the lowest and highest monthly deviation.

[37]In Figure 3.7 I show just three examples of species trait shifts arising from a change in average values of climatic variables, but more complicated effects of increased CO_2 also occur because of changes in the *variation* in, say, temperature (Seneviratne et al. 2006). Consider, for example, that when flowers open earlier a person might first suppose that there could be more abundant crops given a longer growing season, but these early-opening flowers might be more susceptible to late-spring freezes, resulting in more crop failures.

[38]Dunn and Winkler (1999) examine the effects of global warming on the tree swallow laying date.

[39]Schwartz et al. (2006) studied the leaf dates in lilacs and the last freeze dates in the northern hemisphere. In this study the last spring freeze date is defined as the last date with −2.2°C.

[40]Roetzer et al. (2000) demonstrate advancing flowering times in European plants.

(Figure 3.8, p. 82) Species have different features in urban and rural environments.

[41]DeLucia et al. (2005) examined the effect of enhanced CO_2 levels on the biomass produced by trees. I sometimes refer to NPP by a simpler term, *plant growth*, but I mean the same thing. One caution is that there's a difference between how much carbon a plant pulls out of the air and how much change there is in its total mass. Roughly speaking, twice a plant pulls out its NPP from the atmosphere, but respires an amount equal to its NPP, for a net change of its NPP.

[42]Hofmockel and Schlesinger (2007) discuss nitrogen availability to trees under scenarios with enhanced CO_2.

[43]Limitation is an ecological concept: If a rate or density changes when a factor is changed, then that factor can be called "limiting."

[44]In their pollen studies, Ziska et al. (2003) report increased CO_2 levels and temperatures around urban Baltimore, Maryland.

[45]Roetzer et al. (2000) demonstrate advanced flowering times in urban plants.

[46]Neil and Wu (2006) discuss some negative consequences of flowering problems with global warming.

[47]Three examples of changed animal behaviors in urban settings come from Melbourne, Australia. One example shows that the urban climate resulted in a bat, the gray-headed flying fox, establishing new winter camps far outside its usual climatic areas (Parris and Hazell 2005), and two examples demonstrate how urban traffic noise changed the vocalization frequencies of frogs (Parris et al. 2009) and birds (Parris and Schneider 2008).

(Figure 3.9, p. 84) Soils contribute to carbon budgets.

[48]Kaye et al. (2005) compare urban, rural, and native soil respiration. Data points are averages of 6 or 12 samples taken within each month. Pouyat et al. (2006) report urban soil carbon content varying from 1.5 to 16.3 kgC/m^2.

[49]Temporal development of urban soils in Idaho and Washington is described in Scharenbroch et al. (2005). My use of microbial activity paraphrases their technical term, *metabolic quotient*. Their units involve respiration from soil samples and estimates of relative microbial carbon content, two much more complicated measurements than what my simple phrase implies.

[50]Lal (2003) discusses respiration and sequestration of carbon in soils.

(Figure 3.10, p. 86) Vegetation stores and sequesters carbon.

[51]Nowak (1994) discusses carbon sequestration by urban trees.

[52]Jo and McPherson (1995) provide a detailed analysis of the sequestration benefits and maintenance costs of urban vegetation.

[53]Qian and Follett (2002) report carbon sequestration rates in golf courses.

(Figure 3.11, p. 88) Urban pruning can be very intensive.

[54]An example of a case report regarding a tree-trimming death, from www.cdc.gov/niosh/92-106.html: "On October 9, 1990, a 27-year-old tree trimmer (the victim) was working as part of a four-man crew to remove dead trees from a private home in a semi-rural area. The crew had been on the site for 2 days and had removed four large trees. They were working on the fifth tree and had cut off all of the tree limbs. Even though each tree was checked for rot by tapping on the trunk, the crew was unaware of the presence of wood wasps in the upper trunk–a sign of rot. At midmorning, the victim (wearing a saddle belt attached to a cloth lanyard) climbed the tree to cut it away in sections. While ascending the tree, the victim reportedly realized that the tree was more damaged than expected, stopped climbing at approximately 35 feet, and tied off at that height. The rotted tree had a list of approximately 10 to 15 degrees, and as the top section was cut away, the tree bent with it. As the tree sprang back to its original position, the backlash was strong enough to fracture the trunk 6 feet below the cut area where the victim had tied his lanyard. The tree trimmer died when he fell to the ground with the tree section, which landed on top of him."

(Figure 3.12, p. 90) Carbon costs of landscaping machines are high.

[55]Nowak et al. (2002b) discuss carbon costs of power tools and breakeven times for different trees.

(Figure 3.13, p. 92) Durham citizens export their carbon sequestration.

[56]Carbon offset calculations have an element of nonreality. These calculations outrageously assume that the carbon dioxide pulled out of the atmosphere and fixed by vegetation gets locked away, never to be seen again by the biosphere. That idea is not true. Vegetation dies and decomposes, and decomposition releases the carbon dioxide back into the atmosphere. Fossil carbon release represents new carbon added to the biosphere — actually, mainly the atmosphere. I don't

worry too much about making sure these calculations are perfectly precise: People's direct and indirect energy use differs greatly, vegetation patterns vary greatly, and even the sequestration point I've made several times makes me cautious about the underlying offset calculation assumptions.

[57]Price et al. (2002) discuss emissions for electricity generation, and provide an emissions range for electricity generation of about 0.04–0.13 kgC/kWh.

[58]Not every acre of Durham County has trees: See urbanized Durham in Figure 2.1.

[59]Pan et al. (2006) estimated carbon sequestration in northeastern forests.

[60]Cleveland et al. (1999) estimated carbon sequestration in eastern and northeastern forests, and arid shrublands like the southwestern United States.

[61]Another way to think about sequestration is that the 2,300 kg of personally used carbon emissions represents the carbon fixed by about 115 30–45 cm DBH trees, meaning our energy use is equivalent to cutting down about two of these trees each week.

[62]The numbers I report in Figure 3.13 greatly underestimate total U.S. per capita carbon emissions. Dividing total 2008 energy use from Figure 3.1, about 100 quadrillion Btus, by the U.S. population of 300 million leads to 3.3×10^8 Btu/American. Dividing this amount by the 14,000 Btus in a kilogram of wood yields 24,000 kg wood equivalents. That's about half carbon, meaning about 12,000 kg carbon emissions per American, much more than the 2,300 kg in my carbon budget table. Commercial, industrial, and transportation energy use makes up the difference.

[63]I once saw a Web posting announcing that the United Nations had planted more than a billion trees in 2007. That's one tree for each of 7.7 people. It helps, of course, but the problem of increasing carbon dioxide greatly exceeds that solution. Worse yet, even ecological society meetings herald being carbon neutral because organizers paid someone to plant thousands of seedlings, then claimed they offset carbon emissions for a meeting today with carbon sequestration tomorrow. That sounds very much like deficit spending arguments, not true carbon neutrality, from people who should know better, yet chose to fly to attend an international meeting. These tree plantings are a start, of course, but the only real solution involves finding and using fossil-carbon-free energy sources. Planting trees cannot offset our massive fossil-fuel energy consumption.

(Figure 3.14, p. 94) Trees and white paint reduce energy consumption.

[64]I counted 43 tree service entries in the Durham/Chapel Hill phone book. Suppose that each one takes down 200 trees per year, and we're at about 10,000 trees removed in an area with, say, 100,000 houses, or one tree taken down every decade for each house. That's a believable number that agrees with my personal experience (but I have my own chainsaw for the easy situations). If tree removal costs run about $500, that roughly equals the savings on the electric bill over a decade. Motivating trees through summer energy savings doesn't seem very compelling, again due to the tree maintenance issue.

[65]Interestingly, examination of the air conditioning systems in the strip mall that Parker et al. (1997) studied uncovered numerous problems that also led to excess energy use, though repairs were undertaken before the experiment began.

[66]House energy savings through albedo changes are discussed in Akbari et al. (2001), and Parker et al. (2003) have house roof comparisons. For example, nine Florida homes with newly applied high albedo roof coatings, discussed by Akbari et al. (2001), citing Parker (1995), reduced average air conditioning costs by 19%.

[67]Electrical use changes with albedo changes and tree planting in Los Angeles are discussed by Hall (1998).

[68]Cold-climate roof-coating energy issues are discussed in Akbari and Konopacki (2004).

[69]Akbari (2002) discusses indirect sequestration by trees through their energy reduction in Los Angeles and other cities.

[70]General energy reduction strategies are discussed in Akbari and Konopacki (2004).

(Figure 3.15, p. 96) Trees help small houses keep cool and break even for heating.

[71]Simpson (1998) examined trees around buildings in Sacramento, California.

[72]Akbari (2002) describes manipulating trees around houses to study energy savings.

[73]Akbari (2002) cites Parker (1981) for energy savings for trailer houses.

[74]Taha et al. (1991) studied temperature reductions in Davis, California.

[75]Huang et al. (1987) examine many features of house-level energy savings, including insulation.

[76]How many cities have such a database? Could the costs of adding another city employee for this type of study pay for itself in terms of reduced energy and health costs across a city?

(Figure 3.16, p. 98) Wood has low energy content.

[77]My conversions assume 14,000 Btu/kg and 4.1 kWh/kg, and for comparison, gasoline has about 115,000 Btu/gallon, or about 34 kWh/gallon.

[78]Pressing these wood–electricity connections further, a 100-Watt incandescent light bulb burning for 10 hours uses the energy in six-tenths of a pound of firewood.

[79]DeWald et al. (2005) wrote the University of Nebraska Extension Agency guide on wood heating. It also provides nice safety tips and instructions to follow when growing and felling trees for firewood.

[80]Put your energy use into the context of tree biomass. Look outside your window into your own yard and find a dozen 9-inch-diameter trees for each member of your household. Imagine cutting them all down. That's your household's electricity use for the year. Now think about next year's electricity use. Don't even think about your household's gasoline use. Can we really sustain our present energy use with biofuel-based sources?

[81]The British Columbia Ministry of Water, Land and Air Protection (2004) conducted a 2003 survey of household wood use across the province and found that wood-burning households used from 1.6 to 4.9 cords of wood each year.

[82]Houck et al. (1998) discuss the use of various energy sources for heating.

[83]As part of a broader strategy to reduce fossil carbon emissions and carbon dioxide levels, Houck et al. (1998) points out that harvesting older trees for heating and letting younger trees replace them, it can be argued, shifts forests to younger trees and enhances short-term carbon sequestration, as implied in Figure 2.14.

[84]Graham et al. (1992) consider the sequestration rates in short-rotation woody crops (SRWC). Compare this cited sequestration number, 2.2 kg/m^2, to estimates of net carbon sequestration values in cities, calculated by Nowak and Crane (2002), ranging from 0.26 kgC/m^2/year for Atlanta, Georgia, to 0.12 kgC/m^2/year for New York City. Given the desire to yield 2.2 kg/m^2, it's clear we won't grow fuel crops in cities. Compare these numbers to plant growth rates in natural systems, shown in Figure 1.2. We're talking high-growth rates needed for biofuels. Advocates of biomass energy hope to get to 35% efficiency. Corn-to-ethanol conversion has one big strike against it, primarily the fossil-fuel inputs to corn production, but SRWC have less fossil-fuel demand and thus some potential to offset fossil fuels.

[85]Fossil-carbon-free only if one counts any fossil-fuel production use against the biomass's

primary function, not its waste-stream use.

(Figure 3.17, p. 100) Durham citizens use more energy than local forests can provide.

[86]Kiehl and Trenberth (2002) provide a solar radiation budget I use to compare human energy use. These comparisons change pretty drastically for a place like New York City. There citizens have about 40 m^2/person, and if they used the same energy on a daily basis, their energy use would amount to 6,800 Btu/m^2/day. Still, that number is less than the solar input of 16,400 Btu/m^2/day at Earth's surface, but an astonishingly sizable fraction.

[87]I'll point out a serious underestimate here. I used per capita personal use of fuel and electricity, not the U.S.-wide total energy consumption. The latter number means we need to harvest all the biomass off of 5.5 acres per Durham citizen.

[88]This is not possible.

[89]Using the annual sequestration rates for SRWC of 2.2 kg/m^2 mentioned by Graham et al. (1992), use an energy conversion of 14,000 Btu/kg and 365 days per year: (2.2 kg/m^2) (14,000 Btu/kg)/(365 days/year) = 84 Btu/m^2/day.

[90]If everyone in the world used energy like Americans, measured from the U.S.-wide total energy consumption, humans' energy use would equal the entire biosphere's biomass fixation. Biomass is not a viable complete energy solution.

[91]One of the most recent estimates determined that humans capture, most likely, 40% of Earth's net primary productivity — meaning that almost one-half of all the material that plants generate from photosynthesis, either directly or through higher trophic levels. Rojstaczer et al. (2001) updates the Vitousek et al. (1986) calculations that estimated 20 to 40% of the world's primary productivity is captured and used by humans. That's a big, unsustainable ecological footprint for our species. Indeed, accounting for human use of a variety of land and ocean areas, it has been estimated that humans overshot sustainability sometime around 1979, and, in 1999, used about 20% more than the Earth could produce sustainably. Wackernagel et al. (2002) discuss how much humans have overshot sustainability.

Chapter 4: Emissions and Urban Air

(Figure 4.1, p. 104) Human sources of volatile organic compounds (VOCs) are high.

[1]The human point and nonpoint VOC source images from the U.S. EPA were produced from a utility at www.epa.gov.

[2]One document, for example, on Louisiana's Emission Inventory defines point sources as "stationary commercial or industrial operations that emit 100 tons or more per year of VOC or NO_x."

[3]Maantay (2007) examines the connection between point source emissions and minority populations in New York City.

[4]De Gouw et al. (2005) provide an inventory of the volatile organic compounds present in the emissions of fossil fuels in the New England area of North America.

[5]I briefly discussed agricultural nitrogen in Figure 1.6. Nitrogen causes problems, but don't fear all of it. In the form of N_2, it makes up some 78% of our atmosphere. That form of molecular nitrogen has a strong triple bond joining the two atoms together, which few biological processes can tear apart. As a result, that triple bond makes the N_2 molecule nearly irrelevant to issues discussed here.

[6]Olszyna et al. (1997) provide a detailed discussion of nitrogen emissions and ozone production in the southeastern United States.

[7]Evolution of these plumes has been studied mostly during the day when chemical reactions, spurred on by light and atmospheric mixing by winds, take place. See Figure 4.7.

[8]Gillani and Pleim (1996) nicely cover anthropogenic VOC emissions.

(Figure 4.2, p. 106) Fossil-fuel use produces many pollutants.

[9]Emissions data from the U.S. EPA are available at epa.gov/ttn/chief/trends. The U.S. EPA compiled these emissions *estimates* from many sources; the sharp changes arise from new methodologies, measurements, and regulations, not from actual increases or reductions. Likens et al. (2005) discuss some of the mysterious, sharp changes. The numbers I plot for particulate matter include "condensables," which means vapor emissions, not really solid or liquid particles at the moment of emission. This vapor may later condense into very small particles. EPA data also include numbers for particulate matter excluding condensables, showing that about 40% of particulate matter constitutes condensable emissions. In my plot, values for several pollutants must be multiplied by a factor of 10 or 100 (for example, "4" would read "40" or "400").

[10]The EPA website, epa.gov/air/caa, provides an overview of the 1990 Clean Air Act.

[11]I thank an anonymous reviewer for the counterargument that travel doesn't decrease with a service economy, which is certainly true. Yet, I wonder: There *must* be some emissions signature due to this economic shift!

[12]Air pollution measurements in downtown Atlanta during the 1996 Atlanta Summer Olympics are discussed in Friedman et al. (2001). Georgia's Environmental Protection Division monitored air quality before, during, and after the Olympics. The 2008 Beijing Olympics "repeats" these "experiments," but I have not seen any results.

[13]The Clean Air Act of 1990 requires the U.S. Environmental Protection Agency to prepare the National Ambient Air Quality Standard (NAAQS) for many different pollutants.

[14]PM_{10} stands for "particulate matter" smaller than 10 microns — a micron is 10^{-6} meters — really just small particles of junk. Air quality experts now measure even smaller particles, those less than 2.5 microns, which can get trapped deep in the lungs. Their density is denoted by $PM_{2.5}$.

(Figure 4.3, p. 108) Trees produce VOCs.

[15]Lerdau et al. (1997) have a detailed discussion of chemical defenses in trees.

[16]Guenther et al. (1994) discuss emission levels for different tree species. Measurements were taken at the branch level, meaning that they accounted for self-shading and such.

[17]When you think of terpenes, think of *terpentine*, made from the pine tree resin, and menthol as just two examples.

[18]It's neat that natural selection ended up making one molecule, isoprene, useful for one thing (heat protection), and it just so happened that a couple of the molecules stuck together helped protect against herbivory. Evolution just happens, using whatever junk an organism already has built-in. Also, I certainly can't feel ill-will toward trees, what with my use of sunscreens, mosquito and tick repellents, and itch relief compounds, though the natural world might be to blame in the first place. Also note that some VOCs play no functional role in herbivory or stress reduction; these chemicals of life can just leak out from stoma and herbivore wounds.

[19]Geron et al. (2001) argue that numbers like those shown for isoprene and terpene emissions are, in some cases, an order of magnitude too low. Guenther et al. (1994) state that isoprenes and terpenes can account for less than 20% of emissions for some plant species, but can also account for up to 80% for some forests.

[20]Half of full sunlight, which I called average, is considered a photosynthetically active radiation

flux of 1000 μmol/m^2/s, which is nearly equivalent to the heat energy of a 100-Watt bulb every square foot.

[21]President Ronald Reagan's famous 1981 quote that "trees cause more pollution than automobiles do" had a kernel of truth to it, though he ought to have said simply that trees emit volatile organic compounds and left it at that. The "more than automobiles" bit dismisses too much nuance.

(Figure 4.4, p. 110) VOCs produced by trees vary across the contiguous United States.

[22]Guenther et al. (1994) and (2006) estimated isoprene emissions across North America.

[23]For comparison with human emissions, we need to convert the EPA units of tons per square mile per year: (2000 lbs)/(mi^2)/(1 year) =(907,000 g)/(2,600,000 m^2)/(8760 hr) = 4.0×10^{-5} g/m^2/hr = 40 μg/m^2/hr.

[24]Standard conditions are a temperature of 30C and light intensity of 1,000 W/m^2, roughly air temperature and the light energy reaching the ground at the equator on a clear day.

[25]A short note on the EPA website, www.epa.gov/appcdwww/apb/biogenic.htm, written some years ago by Chris Geron, states that anthropogenic VOC emissions are roughly half those of biogenic sources.

[26]Though vegetation might not release much reactive nitrogen, soil-inhabiting organisms and lightning release reactive nitrogen in natural areas (Galloway and Cowling 2002).

[27]Schlesinger and Reiners (1974) showed that a small branch of an artificial tree captured about five times more water and pollutants than an open bucket. More recent approaches place trees in wind tunnels, subject them to salt sprays, and then measure how much salt they capture (Freersmith et al. 2004).

[28]Bolund and Hunhammar (1999) summarize the literature on the physical nature of vegetation filtering out air pollutants. Smith and Jones (2000) provide a current, mechanistic overview of these processes and describe current measurement techniques.

[29]Lovett (1994) studied pollution uptake in tree leaves and also describes a number of artificial surface deposition studies.

[30]Ashmore (2005) provides an excellent discussion of ozone damage to plants, and Karnofsky et al. (2007) give a full review on the consequences for U.S. forests.

(Figure 4.5, p. 112) Trees produce more VOCs in bright light and high heat.

[31]Light intensity in these plots includes only the photosynthetically active radiation (PAR) in the spectrum, usually that within the visible range 400 to 700 nanometers. Units on these scales are μmol/m^2/s, of which only the μmol needs explanation. Recall from chemistry class, Avogadro's number being 6.02×10^{23}, the definition of the number of atoms or molecules in one mole. One can count photons the same way, but the relevant intensity for PAR sunlight is 1,000 μmol/m^2/s, meaning every second about 600,000 quadrillion photons strike a meter-squared area.

[32]White oak emissions data come from two papers of one collaboration. One paper, Baldocchi et al. (1995), reports data from a meeting presentation published later in Harley et al. (1997) with much greater detailed analyses.

[33]Harley et al. (1997) show values for white oak in the range of those shown for beech.

[34]The data are certainly real, but some of the curve-fitting, that I've reproduced here, might be taken with some hesitancy; there's very little data to support the drop-offs at high temperature, even though it's clear that such a drop must happen simply because trees must shut down at too high of a temperature.

[35]Terpenes are mostly carbon by mass, so the slight difference in units, μgC/g/hr versus μg/g/hr, represents just a factor of 0.88 converting emissions per gram from milligrams isoprene to milligrams carbon.

(Figure 4.6, p. 114) VOC sources vary in place and time.

[36]VOC sources changing during the day were disussed by Chameides et al. (1992) and Sillman (1999).

[37]Estimates of the two human source contributions reflect one another in Figure 4.6, at least in part because some VOCs were just divided equally between mobile and stationary.

[38]A description of VOCs can be found in Warneke et al. (2007).

(Figure 4.7, p. 116) VOCs, reactive nitrogen, and sunlight lead to ground-level ozone.

[39]Many atmospheric chemistry books exist; I found the one by Seinfeld and Pandis (2006) helpful.

[40]Though I list CH_3O, the methoxy radical, as the final product of this example using the VOC methane, formaldehyde, CH_2O, actually constitutes an intermediate result, which lasts about four hours in the atmosphere, producing carbon monoxide, CO, which lasts several months, ultimately ending up as carbon dioxide, CO_2. I skipped these steps because many other chemicals get involved, and the figure gets *really* messy.

[41]Baldocchi et al. (1995) provide another full description of ozone formation in the presence of VOCs.

[42]A nanometer (nm) equals 10^{-9} meters. Visible light has wavelengths ranging from 400 to 700 nm, whereas, for example, radio waves (another form of light) have wavelengths of a meter to thousands of meters and very low energy per photon.

[43]The common mnemonic to remember and distinguish frequencies and wavelengths of the visible light spectrum is ROY G. BIV, for red, orange, yellow, green, blue, indigo, and violet. Further, the letter "V" looks like the greek letter ν, the symbol for the frequency of light. Because the V is at the mnemonic's end, violet has the highest frequency. Ultraviolet has an even higher frequency, and infrared — below red — has a low frequency. The energy of a photon of light is proportional to its frequency, and it's the ultraviolet light with high frequency that causes damage. The speed of light, the product of the frequency and the wavelength, is the same for the entire spectrum. Thus, high-frequency ultraviolet light has a short wavelength, and low-frequency infrared light has a long wavelength.

(Figure 4.8, p. 118) Large pollution inputs lead to high downwind ozone levels later.

[44]I thank Noor Gillani for providing the original curves of power plant plumes from the Gillani and Wu (2003) report, which I have not seen. A much earlier, but similar, study with regards to the plume study is Gillani et al. (1981). Other interesting plume studies include Ryerson et al. (2001) and Luria et al. (2003).

[45]Although this graph depicts nitrogen levels, some earlier studies measured sulfur dioxide, SO_2, as a proxy for NO_x levels because it was technically easier. Even though both are major emissions of coal-burning power plants, their chemistries differ, and nitrogen provides the needed information. Connections between SO_2 and NO_x levels have been examined, to some extent, by Likens et al. (2005).

[46]Kuttler and Strassburger (1999) measured ozone and nitrogen levels in Essen, Germany.

[47]Swartz et al. (2005) discuss ozone levels decreasing since 1985, though, in response, Kinney et al. (2005) point out that ozone levels aren't all that responsive to decreasing emissions.

[48]Data from the Bureau of Economic Analysis shows that the U.S. economy has a greater emphasis on producing services, not goods, and the shift occurred sometime just before 1970. Industry-specific contributions to GDP were compiled by the Bureau of Economic Analysis and can be found at www.bea.gov/national/nipaweb. Since about 1950, about 20% of GDP shifted from private goods-producing industries to private services-producing industries. However, as I write this book in the fall of 2008, the global financial world seems to have collapsed. Another couple of years might see a drop-off in the contributions by finance and real estate. Think of these economic data as "snapshots."

[49]Knowlton et al. (2004) predict greater ozone issues with climate change.

[50]Downwind distance changes in ozone production were pointed out by Olszyna et al. (1997).

(Figure 4.9, p. 120) Ozone production and levels have a complicated emissions dependence.

[51]The plot showing how ozone production depends on VOCs and NO_x was adapted from Sillman (1999), which has an excellent discussion of ozone chemistry. The plot depicts results from model calculations discussed therein. The book by Seinfeld and Pandis (2006) and an article by Chameides et al. (1992) also provide good backgrounds.

[52]Kuttler and Strassburger (1999) measured ozone and nitrogen levels in different parts of the city of Essen, Germany.

(Figure 4.10, p. 122) High ozone levels seen Wednesday through Saturday, March through September.

[53]Pollution levels varying with day of the week comes from Cerveny and Balling (1998).

[54]Original data for the weekly ozone-CO levels at Sable Island comes from Parrish et al. (1993).

[55]Pollutants in Boston, Massachusetts were published by Zanobetti and Schwartz (2006).

[56]Nontraffic particulate matter was measured from the residuals of the $PM_{2.5}$ measurements regressed against black carbon; hence, negative values can arise.

(Figure 4.11, p. 124) Ozone in rural areas increases with temperature and nitrogen.

[57]Ozone levels in rural Tennessee were studied by Olszyna et al. (1997).

[58]Galloway and Cowling (2002) discuss various sources of reactive nitrogen, including estimates for natural and historical anthropogenic sources.

[59]Bernard et al. (2001) results suggest what might happen with the increasing temperatures expected with climate change.

[60]I've already shown one example of urban ozone levels from Boston, reported in Figure 4.10: Summer levels reach only about 35 ppb, though the EPA website shows that the Boston area has often exceeded the 90s, violating regulatory standards. I suspect it's just a matter of averaging levels throughout the day and over several years, but the paper (Zanobetti and Schwartz 2006) makes no mention of how the values, provided by the EPA, were averaged.

(Figure 4.12, p.126) When it's hot, urban ozone levels exceed regulatory allowances.

[61]The dependence of ozone levels and maximum daily temperature in various cities was discussed by Bernard et al. (2001) and Pomerantz et al. (1999). Both sources cite other sources for the data, and neither source states whether or not the measurements are one-hour averages, but by the context, I assume that to be the case.

[62]This information on the primary standards for ozone comes from a page on the EPA's epa.gov website, air/criteria, current as of September 15, 2008. These limits are set "to protect public health, including the health of 'sensitive' populations such as asthmatics, children, and the elderly."

Eight-hour standard set in 2008: 0.075 ppm. "To attain this standard, the 3-year average of the fourth-highest daily maximum 8-hour average ozone concentrations measured at each monitor within an area over each year must not exceed 0.075 ppm (effective May 27, 2008)."

Eight-hour standard set in 1997: 0.08 ppm. "(a) To attain this standard, the 3-year average of the fourth-highest daily maximum 8-hour average ozone concentrations measured at each monitor within an area over each year must not exceed 0.08 ppm. (b) The 1997 standard — and the implementation rules for that standard — will remain in place for implementation purposes as EPA undertakes rulemaking to address the transition from the 1997 ozone standard to the 2008 ozone standard."

One-hour standards being revoked: 0.12 ppm. This standard only applies to limited areas. "(a) The standard is attained when the expected number of days per calendar year with maximum hourly average concentrations above 0.12 ppm is < 1. (b) As of June 15, 2005 EPA revoked the 1-hour ozone standard in all areas except the 8-hour ozone nonattainment Early Action Compact (EAC) Areas."

Originally, the one-hour standard was set at 0.08 ppm in 1971, revised to 0.12 ppm in 1979. Whether regulatory or not, the one-hour level still has violations. A quick search found the Texas Commission on Environmental Quality reporting many 2008 summer days exceeding 125 ppb.

[63]White et al. (1994) compare the 1-hour and 8-hour ozone measurements for Atlanta, Georgia.

[64]St. John and Chameides (1997) discuss how really bad ozone days in Atlanta result from specific weather patterns that cook, and recook, the same air mass over several days. They also compare 1-hour and 8-hour averaging.

(Figure 4.13, p. 128) High ozone levels harm vegetation.

[65]Orendovici et al. (2003) provide the ozone damage data for milkweed, *Asclepias incarnata*, and *Viburnum lantana* (called the "Wayfaring tree"), along with reviewing damage to 38 other North American and Spanish shrubs and trees. Pleijel et al. (2007) discuss ozone damage to wheat and potatoes, along with a detailed discussion of stomatal ozone flux.

[66]I found little information regarding the health implications of, say, a 1-hour time span with excessive ozone, while the 8-hour limit is maintained. Making informed regulatory decisions demands information on these potential health implications — if, for example, exceeding some ozone level for five minutes triggers asthma, rather than the exposure integrated over an entire day, then citizens must demand the shorter time scale average limits (see Figure 6.3).

[67]The EPA's website shows continuously updated nonattainment areas across the United States.

(Figure 4.14, p. 130) Air pollution varies greatly in space and time.

[68]Alfani et al. (2005) examined polycyclic aromatic hydrocarbon (PAH) uptake in leaves.

[69]One approach for determining the benefit of trees for air pollution reduction assumes the maximum control cost (MCC) guidelines using nonbiogenic sequestration methods. These guidelines specify maximum cost for cost effectiveness through the "Best Available Control Technology" defined by the U.S. EPA. The Bay Area Air Quality Management District's website, www.baaqmd.gov, has a great deal of information on mitigation issues, and the U.S. EPA New Source Review website, www.epa.gov/nsr, is also informative.

[70]McPherson et al. (1998) present the costs and benefits of trees for reducing air pollution.

[71]In contrast to the actual mechanisms, the graph gives the appearance that green areas might be responsible for high ozone levels. Of course, that's not correct.

[72]McPherson et al. (2005) and McPherson et al. (2006) examine the values of many ecosystem services provided by trees in several cities.

[73]Botanical names for the common species names: dogwood (*Cornus florida*), southern magnolia (*Magnolia grandiflora*), red maple (*Acer rubrum*), and the loblolly pine (*Pinus taeda*).

[74]McDonald et al. (2007) used numerical approaches to estimate the value of tree plantings to mitigate atmospheric particulate matter in two areas of the U.K.

Chapter 5: Social Aspects of Urban Nature

(Figure 5.1, p. 134) What's the value of Chickpea?

[1]Sadly, her name did not protect her from a tragic "accident" involving our dog.

[2]Are we willing to pay for this nonmarket good through increased taxes to support the landowner in his or her retirement, not to mention shoulder the start-up costs for new farmers? In fact, there exist federal programs that buy farmland conservation easements. These programs buy the development rights of farmland from the landowners. This purchase lowers the land's value, provides the landowner some money in return, and sometimes involves a requirement that the land be farmed in perpetuity (see, for example, www.nrcs.usda.gov/programs/frpp).

[3]Hanemann (1994) and Carson et al. (2001) provide introductions to contingent valuation. Another important, detailed example is the economic value of water (Hanemann 2006).

[4]An example of Carson et al. (2001) compares turning a parcel of land into a stripmine versus a closed wildlife reserve.

(Figure 5.2, p. 136) S.A. Forbes (1880) estimates the value of birds.

[5]An excellent review of ecosystem services and valuation was done by Farber et al. (2006).

[6]Guillebeau et al. (2006) report insecticide costs and benefits in Georgia.

[7]Spider bites were also included within the Georgia public health publication (Guillebeau et al. 2006) on insect costs; let's not be pedantic over the term *insects*.

[8]I found a Web-based utility, apparently run by academics, that compares currency values across large spans of time at www.measuringworth.com. It shows that $0.56 from 1880 had the following values in 2003: $10.40 using the Consumer Price Index; $9.77 using the GDP deflator; $72.62 using the unskilled wage; $102.33 using the nominal GDP per capita; and $592.38 using the relative share of GDP. I take that to mean my calculation of $18 per acre for the damage done by insects, and $10 per acre for the value of birds' "ecosystem service," is comparable to Forbes's values.

(Figure 5.3, p. 138) Trees make satisfying neighborhoods.

[9]Correlations near zero mean very little connection, values near ± 1 mean strongly connected, and negative numbers mean an increase in one reflects a decrease in the other. See Figure A.3 for a deeper understanding of correlations.

[10]Ellis et al. (2006) studied people's satisfaction with where they lived and demonstrated that trees "moderated" people's negative feelings associated with living near retail areas.

[11]Carson et al. (2001) argue in favor of evaluating nonmarket values through surveys and simultaneously present a good introduction. Harrison (1992) also critically examines these issues.

[12]An animal-rendering plant processes animal carcasses after butchering and can lead to considerable odors. Bowker and MacDonald (1993) studied WTA and WTP values for the animal-rendering plant, and note many cautions to such an examination.

[13]The question of why WTP values and WTA values are different has led to some impressively mathematical demonstrations that the two numbers should not be identical (Hanemann 1991).

[14]Svedsater (2003) describes detailed responses to interviews that assess these WTA and WTP values and concludes that people come up with these values using many irrelevant, noneconomic thought processes.

[15]Horowitz and McConnell (2002) performed the WTA-to-WTP ratio meta-analyses determining what kinds of goods have the highest ratios.

[16]Instead of buying land outright, groups can purchase the right to develop land into suburbs or what-not, while the land's owner retains all other rights. The land's deed notes the easement and provides legal recourse if future landowners violate the restriction. Farmland conservation easements can also require that the land be used for agriculture in perpetuity.

(Figure 5.4, p. 140) People like neat trees, not messy forests.

[17]Talbot and Kaplan (1984) performed the surveys of liked and disliked features about urban open spaces.

[18]Of course, nature provides negative experiences, too. I once took a walk in my North Carolina woods at the end of August and came back with about 100 two-week-itch-worthy chigger bites. How do I value that aspect of nature?

[19]Anecdotes like urban foresters reporting "mow it down now!" calls based on snake sightings have support in controlled surveys. For example, Knight (2008) reports that, in addition to snakes, people hate wolves, cougars, bats, and spiders, too, though they tend to support preservation of the mammals through government regulation more strongly than the others.

(Figure 5.5, p. 142) Park features involving scenic beauty and perceived security.

[20]Schroeder and Anderson (1984) rated park features according to security and scenic beauty.

(Figure 5.6, p. 144) Underbrush was bad as far back as 1285.

[21]I initially happened upon the 1285 Statute of Winchester quote, in a shortened form, in Kuo and Sullivan (2001a), within which they examine the connections between crime and vegetation. The shortened form makes the statute seem much more severe, but the earlier quote shows deeper concern only with underbrush.

[22]Brundson et al. (1995) discuss some of these fear-of-vegetation issues, disputing simplistic ideas that removing vegetation reduces crime.

[23]A U.S. Bureau of Justice Statistics bulletin by Catalano (2006) provides an excellent summary of crime experienced by Americans.

[24]Kuo and Sullivan (2001a) provide a fascinating overview of studies linking fear of crime and vegetation.

[25]Michael et al. (2001) surveyed informants regarding auto burglaries and the role vegetation played.

(Figure 5.7, p. 146) About 10 out of 10 people prefer malls.

[26]I fear any democratic appeal pitting the environment against new commercial or residential developments: When comparing the popularity of a natural park and a mall, a new commercial

development seemingly wins by a landslide!

[27]Carson et al. (2003) provide background on economic valuation, and a great summary article explaining many issues is Bingham et al. (1995), reporting the results of a 1991 forum of experts in the area sponsored by the U.S. Environmental Protection Agency (EPA).

[28]Bergstrom (1990) reviewed examples of economic valuation.

[29]Costanza et al. (1997) and Rojstaczer et al. (2001) provide thorough coverage of valuing ecosystem services. Many citations to these papers provide full coverage.

[30]Boumans et al. (2002) used an enormous model to estimate that the services provided by ecosystems was about 4.5 times the gross world product in the year 2000.

[31]Costanza and Daly (1992) provide an early discussion of natural capital and sustainability.

[32]An excellent discussion of value, services, and ecological processes is provided in Farber et al. (2002, 2006). Dukes and Mooney (2004) discuss the disruption of ecosystems, including their services, due to invasive exotic species. Once values are determined in one place, transferring values determined from one place and time to the benefits provided by the environment in a different time and place is problematic. Nothing says the values are the same, and Spash and Vatn (2006) provide a discussion of this issue.

[33]The pros and cons of economic valuation are discussed by McCauley (2006), and further discussed by several others in response: Marvier et al. (2006), Reid (2006), and Costanza (2006).

[34]A recent example: A case before the U.S. Supreme Court involved power plant cooling systems that use huge volumes of water for cooling their systems, simultaneously heating the water. This water could come from "closed-loop" reservoirs designed solely for cooling, or large rivers, lakes, or oceans. In the latter case, anything living in the natural waters either gets mashed by the tremendous pressures against screens over the intake pipes, or cooked in the cooling system when the water heats up. Chapter 26 of the U.S. statutes (number 33) on navigable waters, section 1326(a), titled "Effluent limitations that will assure protection and propagation of balanced, indigenous population of shellfish, fish, and wildlife," reads, "With respect to any point source . . . , whenever the owner or operator of any such source . . . can demonstrate to the satisfaction of the Administrator (or, if appropriate, the State) that any effluent limitation proposed for the control of the thermal component of any discharge from such source will require effluent limitations more stringent than necessary to assure the projection and propagation of a balanced, indigenous population of shellfish, fish, and wildlife in and on the body of water into which the discharge is to be made, the Administrator (or, if appropriate, the State) may impose an effluent limitation under such sections for such plant, with respect to the thermal component of such discharge (taking into account the interaction of such thermal component with other pollutants), that will assure the protection and propagation of a balanced, indigenous population of shellfish, fish, and wildlife in and on that body of water." In different legislation, Section 316(b) of the Clean Water Act reads: "Any standard established pursuant to section 301 or section 306 of this Act and applicable to a point source shall require that the location, design, construction, and capacity of cooling water intake structures reflect the best technology available for minimizing adverse environmental impact." On April 1, 2009, the U.S. Supreme Court ruled 6–3 that power plants can cook up aquatic wildlife because the costs of protecting them exceed the benefits. The majority opinion states "that the EPA permissibly relied on cost-benefit analysis in setting the national performance standards." Dissenting, Justice Stevens, with Justices Souter and Ginsburg joining, write: "the EPA estimated that water intake structures kill 3.4 billion fish and shellfish each year," but "instead of monetizing all aquatic life, the Agency counted only those species that are commercially or recreationally

harvested, a tiny slice (1.8 percent to be precise) of all impacted fish and shellfish. This narrow focus in turn skewed the Agency's calculation of benefits. When the EPA attempted to value all aquatic life, the benefits measured $735 million. But when the EPA decided to give zero value to the 98.2 percent of fish not commercially or recreationally harvested, the benefits calculation dropped dramatically — to $83 million." Justice Stevens continues, "The Agency acknowledged that its failure to monetize the other 98.2 percent of affected species 'could result in serious misallocation of resources' because its 'comparison of complete costs and incomplete benefits does not provide an accurate picture of net benefits to society.'" Nature dies on the economic sword.

(Figure 5.8, p. 148) Varying tree cover in Chicago public housing.

[35]Kuo and Sullivan (2001a) report on the crime rate dependence on vegetation in the Ida B. Wells public housing development in Chicago, Illinois. Though the results are significant, vegetation explains only 7 to 8% of the variation in crime rates. Since the Kuo and Sullivan (2001a) study took place, the Chicago Housing Authority (CHA) combined the development with others, presently housing around a couple thousand people. Information on the current status of the Ida B. Wells public housing development can be found on the CHA's website: www.thecha.org/housingdev/-madden_wells.html.

[36]On the one hand, apartment buildings and housing complexes provide a fantastic opportunity to learn how vegetation and its correlates affect people, but on the other hand, a vague, uneasy feeling comes over me about using these people's difficult situation as the basis of a scientific study. I resolve this conflict in favor of learning about the people–vegetation link when I see the socioeconomic inequity in vegetation that I show in Figure 6.13. Furthermore, there's quite a long history involving public housing in the United States, the responsibility of which once rested on families and communities to take care of their old, infirm, and poor. As communities grew into cities, public housing organization grew similarly. Clearly, studying people *is* the nature of the social sciences, and it's a valid question to ask what consequences result when people have very little vegetation.

(Figure 5.9, p. 150) Reduced vegetation correlates with higher crime.

[37]Several important features complicate the connection between crime rates and vegetation. For example, vacancy rate had no effect on crime rates, though the number of units per building was an important factor. Further, although crime rate increased and vegetation decreased with increasing building height, decreasing crime rates remained correlated with increasing vegetation after correction.

[38]Sullivan et al. (2004) study social interactions and their dependence on vegetation levels at the Ida B. Wells buildings. Five landscape architecture and horticulture students determined each of 59 locations' "greenness," classifying 27 locations as barren and 32 as green. Next, researchers working with two residents of a different housing complex, presumably minimizing the impact from observations, observed each of the 59 locations three times on weekdays between 3:30 and 6:15 PM, and once on Sunday between 12:15 and 5:15 PM. Observers coded people's behaviors like sitting, talking, or eating with a group of people as social activities, and activities like reading a book as nonsocial activities. In all, 90% more people used greenspaces than barren spaces, and 83% more engaged in social activities in greenspaces compared to barren spaces. I transformed the reported standard deviations into standard deviations of the means.

[39]Kuo and Sullivan (2001b) report studies of domestic violence in the Robert Taylor Homes development. Independent from the reported incidents of violence, 22 horticultural stu-

dents, both undergraduates and graduate students, rated a building's "greenness" on a scale from 0 (not at all green) to 4 (very green).

[40]Taylor et al. (2002) describe attention restoration theory and cite a number of studies that confirm the idea.

[41]Although valuing crime reduction in economic terms seems vaguely disturbing, the U.S. Department of Justice attempts it, with economic costs of crime summarized in Table 82, Personal and property crimes, 2005, from the Bureau of Justice Statistics, U.S. Department of Justice. As might be expected, the Justice Department finds a rather minimal total, direct economic loss to crime victims. One statistic, in particular, points out the absurdity of estimating the effects of crime in economic terms: The average economic loss to rape victims was $183. Certainly, everyone, including the Department of Justice, understands the devastation that the crime of rape imposes on its victims, but the crime causes lasting devastation to a nonmarket good, the victim's sense of well-being. Further along these lines, environmental economists at the EPA summarized studies from 1976 to 1991 that used contingent valuation to estimate the "statistical value" of a life (Dockins et al. 2004). Values, in 2002 dollars, range from $0.9 million to $20.9 million. The EPA has used values ranging from $5.5 to $6.2 million.

[42]An approach called Crime Prevention Through Environmental Design (CPTED) was developed along these lines. Internet searches yield much information on the topic, but peer-reviewed studies appear to be scant.

(Figure 5.10, p. 152) Girls' self-discipline develops better with nature.

[43]Taylor et al. (2002) studied childhood development in the Taylor Buildings in Chicago, Illinois, and provide a broad review of attention restoration theory. Children are not alone in their benefiting from nature: Benefits also extend to adults in that natural views at one's residence have a restorative effect when one is suffering from fatigue.

[44]In contrast to the studies I described previously in Figure 5.9, this one determined the "greenness" of a view through the interviewees themselves. Two questions were used and combined: "How much of the view from your window is of nature (trees, plants, water)?" and the reverse, "How much of your view from your window is man-made (buildings, street, pavement)?" Responses ranged from 0 = not at all to 4 = very much.

(Figure 5.11, p. 154) Nature promotes emotional and physical health.

[45]Gidlöf-Gunnarsson and Öhrström (2007) conducted the Swedish study regarding the effect of access to green areas on people's feelings. They also examined the importance of having a quiet side to one's apartment, finding that availability similarly important.

[46]Fuller et al. (2007) examined how *reflection* and *distinct identity* depend on species richness in Sheffield, UK. Another interesting test they made shows that untrained people perceive species diversity that is much in line with actual empirical measures of species richness. Bird and butterfly diversity failed to produce similar trends. Significant trends in these emotional measures with plant diversity remained after correction for habitat area, which itself was positively correlated, along with habitat diversity.

[47]Bell et al. (2008) showed the highly significant trend of the body mass index (BMI) with normalized difference vegetation index (NDVI) in the Indianapolis, Indiana, study of nearly 4,000 children aged 3–16.

(Figure 5.12, p. 156) Trees promote bird and plant species richness.

[48]Donnelly and Marzluff (2006) performed an extensive study of bird species richness in Seattle, Washington, and recommend having at least 10 trees per hectare in cities. Hansen et al. (2005) reviewed many aspects of urban sprawl on biodiversity and reported the plant species diversity effects of forest cover.

[49]My piecewise fits to the Donnelly and Marzluff (2006) data may be questionable; indeed the original authors didn't do it, and their statistical analyses may be preferred over my approach. However, the large gap in the middle of tree density causes practical interpretational challenges. For my analysis, the fit for low tree density gives $R^2 = 0.49$ and and a fit $y = 14.1 + 0.164x$. The fit at higher density is almost significant ($p = 0.067$) with a fit $y = 23.2 + 0.020x$.

[50]Birds like to roost in evergreens. I remember once searching for a missing chicken one night, and being startled when parting the branches of a backyard cedar tree, I found a sleeping cardinal.

[51]Haire et al. (2000) examined grassland bird species in Boulder, Colorado, and found decreasing richness with urbanization. They also make recommendations on urban land cover.

[52]McCarthy et al. (2006) discuss sources and sinks for colonization and extinctions.

Chapter 6: Human Health and Urban Inequities

(Figure 6.1, p. 160) Heat waves lead to deaths a few days later.

[1]McGeehin and Mirabelli (2001) discuss many aspects of heatstroke and heat-related mortality, including the feature that mortality takes place a few days after a heat event.

[2]Regarding the time delay between peak heat and peak deaths, I quote one case report from the Centers for Disease Control (CDC 1995): "On June 13, 1994, in Houston, Texas, a 29-year-old mentally impaired women was found lying on the floor of her garage. She was unresponsive when admitted to a local hospital and had a rectal temperature of 107.9F (41.9C). She died within 2 days of arrival at the hospital. The outdoor temperature and humidity had reached 92.0F (33.3C) and 91%, respectively."

[3]The U.S. National Oceanic and Atmospheric Administration (NOAA) provides a very detailed polynomial expansion for the approximation to the dangerous heat index threshold, T, given the relative humidity, R, but I daresay $R = 100 - 6(T - 90)$ is much simpler. Maybe even easier is the series of the first three square numbers, 1–4–9: 100F at 40% and 90F at 100%. NOAA provides a fine description of heat effects and symptoms at www.nws.noaa.gov/om//brochures/heat_wave.shtml

[4]Information comes from a report by the Centers for Disease Control (CDC 1995).

(Figure 6.2, p. 162) Particulate matter is bad for older people.

[5]Zanobetti and Schwartz (2006) studied pollution effects on health in the greater Boston area.

(Figure 6.3, p. 164) High ozone and SO_2 levels predict high asthma hospitalizations.

[6]A myriad of health problems brought on by air pollution are discussed by Bernard et al. (2001).

[7]Lin et al. (2004) discuss mechanisms of asthma inducement.

[8]Byrd and Joad (2006) provide two references on age differences in asthma susceptibility.

[9]Lin et al. (2004) and White et al. (1994) discuss asthma emergencies increasing with pollution levels, SO_2 and ozone, respectively.

[10]A one-day lag, for example, simply means that one day a person experiences high pollutant

levels, and then one day later the hospital admission takes place.

[11]Tolbert et al. (2000) showed the spatial correlations between ozone and asthma admissions in Atlanta, Georgia.

[12]Peel et al. (2005) examined emergency room visits in Atlanta, Georgia, from 1993 to 2000.

[13]The idea that ozone alone accounts mostly for asthma came out of another study from Southern California by Moore et al. (2008). This study also provides maps showing significant reductions in ozone and PM_{10} from 1983 to 2000.

[14]The four database sources included Medicaid claims, a health maintenance organization, combined emergency department records from two pediatric hospitals, and a hospital discharge database from all of Atlanta's metropolitan hospitals.

(Figure 6.4, p. 166) Asthma incidence and pollution aren't tightly correlated through time.

[15]Sun et al. (2006) examined asthma cases in Taiwan.

[16]Hashimoto et al. (2004) discuss a range of environmental features that increase the risk of asthma attacks, including a rapid decrease in temperature, humidity, and pressure.

[17]D'Amato et al. (2007) talk about pollen release, with thunderstorms being a potential cause of asthma attacks.

[18]Byrd and Joad (2006) describe a multitude of urban-related causes of asthma.

(Figure 6.5, p. 168) Though cities differ, heat kills people in July and August.

[19]Kalkstein and Davis (1989) present the mortality versus temperature plots, along with a great deal of further analyses. My digitization of their plots gave around 250 points per plot for the summer months of these 11 years; 11 years times 90 days would be 990 points. I believe the authors plotted each day of the three summer months for each of the 11 years, but many points sit on top of one another (they make no explicit statement clarifying this point in the paper). I dropped an off-axis point for New York City having a standard mortality of 250 at 40C.

[20]Kalkstein and Greene (1997) discuss weather patterns and mortality, as well as heat-related mortality increases with global warming.

[21]Davis et al. (2003, 2004) performed the analysis of heat-related deaths in 28 metropolitan areas.

[22]Anderson and Rosenberg (1998) describe standardized mortality calculations.

[23]Chip Knappenberger, one of the authors of the Davis et al. (2003, 2004) studies, writes, "For our purposes, which were to best isolate a weather component as a cause of mortality, we age-standardized to remove changing population demographic effects from the annual mortality numbers, just as the CDC recommends in order to compare from place to place and from time to time. Once we had best taken care of demographic issues, we still had a big trend to deal with — presumably related to technology improvements (medicine, etc.) as well as a seasonal cycle. We attempted to remove those effects by subtracting the monthly median daily mortality (for a particular month–year) in order to arrive at daily mortality anomalies, which we then related to weather variables (specifically 1600LST apparent temperature)."

(Figure 6.6, p. 170) Air conditioning reduces heat-related mortality.

[24]Electricity use data with temperature in Figure 6.6 was provided by Guido Franco, discussed in Franco and Sanstad (2006). The data come from CalISO, the electric company that serves much of California.

[25]Akbari et al. (2001) estimated the energy use due to the UHI effect.

[26]Davis et al. (2003) show heat-related mortality reductions through air conditioning use.

[27]Pomerantz et al. (1999) estimated U.S. air conditioning costs.

(Figure 6.7, p. 172) Many people die in winter (accounting for age, race, and gender).

[28]The standardized mortality data shown in Figure 6.7 was kindly provided by Chip Knappenberger, a coauthor of the Davis et al. (2003, 2004) work. See Figure 6.5's notes for further discussion.

[29]I found age-specific mortality data at the CDC's National Center for Health Statistics website, specifically Worktable 310, www.cdc.gov/nchs/datawh/statab/unpubd/mortabs/gmwk310_10.htm. I found the population estimates for each age, sex, and race cohort from the U.S. Census site, www.census.gov/popest/national/asrh/2007-nat-res.html. My results simply divide the number of deaths by the population.

[30]As one important example, countries go through "demographic transitions" in which a developing economy experiences decreasing mortality rates while maintaining high birth rates. As a result, the population rapidly increases with many more younger people. As the country develops later, the birth rate declines, the pulse of young people age, and the contracting population has relatively more older people. Average mortality rate changes throughout this period, and uncovering the effects of heat and air pollution out from under the stronger age-dependent signal must be difficult.

(Figure 6.8, p. 174) Lower income, fewer trees, and higher temperatures go together in Durham.

[31]Mennis (2006) examined vegetation and socioeconomic variables in Denver, Colorado, using two different approaches: a standard multivariate regression and a novel rule-mining approach.

[32]Thompson (2002) states that changing the "pattern" of a city from a centralized one to compact neighborhoods or urban areas with hubs helps equity, and outlines several reasons to have this space available.

(Figure 6.9, p. 176) Wealth, homeownership, and trees connect in Milwaukee, Wisconsin.

[33]Heynen et al. (2006) examine canopy cover in Milwaukee, Wisconsin.

[34]Milwaukee's forestry department manager statement as expressed in a footnote of Heynen et al. (2006).

[35]Perkins et al. (2004) discuss the inherent inequity in canopy cover resulting from tree-planting programs that focus on owner-occupied dwellings in the face of canopy deficits in high-rentership areas of Milwaukee.

(Figure 6.10, p. 178) Wealth, education, and vegetation correlate in Baltimore, Maryland.

[36]Grove and Burch (1997) studied vegetation and socioeconomic variables in Baltimore, Maryland. This work has been extended and updated by Grove et al. (2006) and Troy et al. (2007).

(Figure 6.11, p. 180) Parks, trees, and plants come with wealth.

[37]In a personal communication, Jennifer Wolch, the lead author of the Wolch et al. (2005) Los Angeles study, described the unique areas serviced by parks as "Theissen polygons." A park's polygon includes all locations that have that park as its closest one. Edges of adjoining polygons represent locations equidistant from two or more parks. These polygons cover the entire study area.

[38]The fit of median income and tree canopy across Durham County is significant with $R^2 = 0.45$ and linear fit of form, $y = 11.24 + 0.000696x$.

[39]Hope et al. (2003) present the inequitable distribution of plant genera against family income in Phoenix, Arizona.

(Figure 6.12, p. 182) Minority populations have worse air, income, and asthma.

[40]Sister et al. (2007) connected the park provisioning to racial distributions in Los Angeles, California. A park's service area was determined as discussed in a note for Figure 6.11.

[41]Maantay (2007) shows racial and poverty inequities in New York City.

[42]Downey (2007) examines racial inequities in income and air quality in MSAs across the United States.

(Figure 6.13, p. 184) Healthier neighborhoods are usually wealthier neighborhoods.

[43]Harlan et al. (2006) examined the heat comfort index with average income in Phoenix, Arizona.

[44]See Figure A.6 for a discussion of the differences between standard deviations and standard errors.

[45]Mitchell and Dorling (2003) studied NO_2 pollution over all wards in Great Britain. This fascinating study looked at many different features concerning who pollutes and who breathes the pollution.

[46]Mitchell and Popham (2007) studied health across England.

(Figure 6.14, p. 186) Income helps education and increases life expectancy.

[47]Unfortunately, students themselves reported these incomes. Perhaps, kids doing poorly on the SATs might preferentially underestimate family incomes, and students doing well might overestimate family income. Of course, I can also imagine the exact opposite situation.

[48]The College Board report from which this SAT–income data originated was unclear whether the reported standard deviation of around 100 points was with respect to the distribution or the mean, but given the SAT scores, 100 points certainly looks like a reasonable value for the standard deviation. To understand the trends better, I have assumed so and calculated the standard deviation of the means — also called the standard error — which is much smaller, roughly 2 points, smaller than the plotting symbols.

[49]The data show no connection between percentage of students tested and combined SAT scores at the county level. The fit of 2005 SAT scores and 2004 per capita income is significant with $R^2 = 0.21$ and linear fit of form, $y = 787 + 0.00746x$.

[50]Ezzati et al. (2008) provide county-level life expectancy data and show increased disparity in life expectancy with economic status.

[51]The fits for life expectancy versus income in Figure 6.14, is significant with $R^2 = 0.24$; fitting the functional form, $y = 70.0 + 0.00020x$.

[52]The fits for life expectancy versus SAT score in Figure 6.14 is significant with $R^2 = 0.57$; fitting the functional form, $y = 56.3 + 0.019x$.

Appendix: Graphical Intuitions

(Figure A.1, p. 196) Three equivalent data representations.

[1]Some people call the horizontal axis the abscissa and the vertical axis the ordinate, or at least, by definition, the distances of points along each axis. I don't use these terms, and I rarely hear other

scientists use them. I must admit I've always found the two terms confusing, just like the terms stalagmites and stalactites. I really prefer calling the axes by x and y, harking back to my training in the physical sciences. The most common custom, however, places the "independent" variable on the horizontal axis and the "dependent" variable on the vertical axis.

(Figure A.3, p. 200) Importance versus significance.

[2]Statisticians recognize a difference between R^2 and r^2, one involving situations with a single variable and the other with multiple variables. I make no such distinction.

[3]Shameless self-promotional material: Learn all about C-programming applied to ecological and evolutionary problems in Wilson (2000).

[4]Vociferous arguments abound regarding the utility and misinterpretations of the p-value. I won't go into those arguments, but look at its measure of significance as a statistical guide, not a hard-and-fast rule.

(Figure A.6, p. 206) Sample sizes and measures of variation.

[5]Normal distributions are a more technical name for the "bell curves" sometimes mentioned more informally. They also go by the name Gaussian distribution. I could show the formula, but more complete descriptions can be found elsewhere because they're so normal.

[6]I gave both distributions the same standard deviation, 0.1, meaning roughly 68% of the points will fall between 0.39 (0.41) and 0.59 (0.61). This is verified by the really dark region in those bands for the high sample size example.

References

Adams, G.B., and H.M. Stephens. 1901. *Select Documents of English Constitutional History.* Macmillan & Co., Ltd.

Akbari, H. 2002. Shade trees reduce building energy use and CO_2 emissions from power plants. *Environmental Pollution* 116 (2002): S119–S126.

Akbari, H., and S. Konopacki. 2004. Energy effects of heat-island reduction strategies in Toronto, Canada. *Energy* 29: 191–210.

Akbari, H., M. Pomerantz, and H. Taha. 2001. Cool surfaces and shade trees to reduce energy use and improve air quality in urban areas. *Solar Energy* 70: 295–310.

Al-Abdul Wahhab, H.I., and F.A. Balghunaim. 1994. Asphalt pavement temperature-related to arid Saudi environment. *Journal of Materials in Civil Engineering* 6: 1–14.

Alberti, M., J.M. Marzluff, E. Shulenberger, G. Bradley, C. Ryan, and C. Zumbrunnen. 2003. Integrating humans into ecology: Opportunities and challenges for studying urban ecosystems. *Bioscience* 53: 1169–1179.

Alfani, A., F. De Nicola, G. Maisto, and M.V. Prati. 2005. Long-term PAH accumulation after bud break in *Quercus ilex* L. leaves in a polluted environment. *Atmospheric Environment* 39: 307–314.

Allan, B.F., F. Keesing, and R.S. Ostfeld. 2003. Effect of forest fragmentation on Lyme disease risk. *Conservation Biology* 17: 267–272.

Anderson R.N., and H.M. Rosenberg. 1998. Age standardization of death rates: Implementation of the year 2000 standard. National Vital Statistics Reports, Vol. 47, No. 3. Hyattsville, MD: National Center for Health Statistics.

Annual Energy Review. 2008. U.S. Department of Energy. DOE/EIA-0384.

Arnfield, A.J. 2003. Two decades of urban climate research: A review of turbulence, exchanges of energy and water, and the urban heat island. *International Journal of Climatology* 23: 1–26.

Ashmore, M.R. 2005. Assessing the future global impacts of ozone on vegetation. *Plant, Cell and Environment* 28: 949–964.

Ashworth, J.R. 1929. The influence of smoke and hot gases from factory chimneys on rainfall. *Quarterly Journal of the Meteorological Society* 55: 341–350.

Ashworth, J.R. 1944. Smoke and rain. *Nature* 154: 213–214.

Auerbach, B.M., and C.B. Ruff. 2004. Human body mass estimation: A comparison of "morphometric" and "mechanical" methods. *American Journal of Physical Anthropology* 125: 331–342.

Badarinath, K.V.S., T.R. Kiran Chand, K. Madhavi Latha, and V. Raghavaswamy. 2005. Studies on urban heat islands using ENVISAT AATSR data. *Journal of the Indian Society of Remote Sensing* 33: 495–501.

Baldocchi, D., A. Guenther, P. Harley, L. Klinger, P. Zimmerman, B. Lamb, and H. Westberg. 1995. The fluxes and air chemistry of isoprene above a deciduous hardwood forest. *Philosophical Transactions: Physical Sciences and Engineering* 351: 279–296.

Balmford, A., A. Bruner, P. Cooper, R. Costanza, S. Farber, R.E. Green, M. Jenkins, P. Jefferiss, V. Jessamy, J. Madden, K. Munro, N. Myers, S. Naeem, J. Paavola, M. Rayment, S. Rosendo, J. Roughgarden, K. Trumper, and R.K. Turner. 2002. Economic reasons for conserving wild nature. *Science* 297: 950–953.

Bell, T.L., D. Rosenfeld, K.-M. Kim, J.-M. Yoo, M.-I. Lee, and M. Hahnenberger. 2008. Midweek increase in U.S. summer rain and storm heights suggests air pollution invigorates rainstorms. *Journal of Geophysical Research* 113: D02209.

Bergstrom, J.C. 1990. Concepts and measures of the economic value of environmental quality: A review. *Journal of Environmental Management* 31: 215–228.

Bernard, S.M., J.M. Samet, A. Grambsch, K.L. Ebi, and I. Romieu. 2001. The potential impacts of climate variability and change on air pollution-related health effects in the United States. *Environmental Health Perspectives* 109: 199–209.

Bingham, G., R. Bishop, M. Brody, D. Bromley, E. Clark, W. Cooper, R. Costanza, T. Hale, G. Hayden, S. Kellert, R. Norgaard, B. Norton, J. Payne, C. Russell, and G. Suter. 1995. Issues in ecosystem valuation: Improving information for decision making. *Ecological Economics* 14: 73–90.

Bock, C.E., Z.F. Jones, and J.H. Bock. 2008. The oasis effect: Response of birds to exurban development in a southwestern savanna. *Ecological Applications* 18: 1093–1106.

Bolund, P., and S. Hunhammar. 1999. Ecosystem services in urban areas. *Ecological Economics* 29: 293–301.

Borenstein, S. 2008. The market value and cost of solar photovoltaic electricity production. Center for the Study of Energy Markets Working Paper 176 (www.ucei.org).

Bornstein, R., and Q. Lin. 2000. Urban heat islands and summertime convective thunderstorms in Atlanta: Three case studies. *Atmospheric Environment* 34: 507–516.

Boumans, R., R. Costanza, J. Farley, M.A. Wilson, R. Portela, J. Rotmans, F. Villa, and M. Grasso. 2002. Modeling the dynamics of the integrated earth system and the value of global ecosystem services using the GUMBO model. *Ecological Economics* 41: 529–560.

Bowker, J.M., and H.F. MacDonald. 1993. An economic analysis of localized pollution: rendering emissions in a residential setting. *Canadian Journal of Agricultural Economics* 41: 45–59.

Brattebo, B.O., and D.B. Booth. 2003. Long-term stormwater quantity and quality performance of permeable pavement systems. *Water Research* 37: 4369–4376.

British Columbia Ministry of Water, Land and Air Protection. 2004. Residential wood burning emissions in British Columbia. ISBN 0-7726-5438-7.

Brohan, P., J.J. Kennedy, I. Harris, S.F.B. Tett, and P.D. Jones. 2006. Uncertainty estimates in regional and global observed temperature changes: A new data set from 1850. *Journal of Geophysical Research* 111: D12106.

Brunsdon, C., R. Gilroy, A.M. Pour, M. Roe, I. Thompson, and T. Townshend. 1995. Safety, crime, vulnerability and design: A proposed agenda of study. School of Architecture, Planning & Landscape: Global Urban Research Unit. Working Paper No. 53. University of Newcastle upon Tyne.

Burian, S.J., and J.M. Shepherd. 2005. Effect of urbanization on the diurnal rainfall pattern in Houston. *Hydrological Processes* 19: 1089–1103.

Byrd, R.S., and J.P. Joad. 2006. Urban asthma. *Current Opinion in Pulmonary Medicine* 12: 68–74.

Carle, M.V., P.N. Halpin, and C.A. Stow. 2005. Patterns of watershed urbanization and impacts on water quality. *Journal of American Water Resources Association* 41: 693–708.

Carreiro, M.M., and C.E. Tripler. 2005. Forest remnants along urban-rural gradients: Examining their potential for global change research. *Ecosystems* 8: 568–582.

Carrow, R.N. 1995. Drought resistance aspects of turfgrasses in the Southeast: Evapotranspiration and crop coefficients. *Crop Science* 35: 1685–1690.

Carson, R.T., N.E. Flores, and N.F. Meade. 2001. Contingent valuation: Controversies and evidence. *Environmental and Resource Economics* 19: 173–210.

Carson, R.T., R.C. Mitchell, M. Hanemann, R.J. Kopp, S. Presser, and P.A. Ruud. 2003. Contingent valuation and lost passive use: Damages from the *Exxon Valdez* oil spill. *Environmental and Resource Economics* 25: 257–286.

Carver, A.D., D.R. Unger, and C.L. Parks. 2004. Modeling energy savings from urban shade trees: An assessment of the CITYgreen energy conservation module. *Environmental Management* 34: 650–655.

Catalano, S.M. 2006. *Criminal Victimization, 2005.* U.S. Bureau of Justice Statistics Bulletin NCJ 214644.

Centers for Disease Control. 1995. Heat-related illnesses and deaths — United States, 1994–1995. *Morbidity and Mortality Weekly Report* 44(25): 465–468.

Cerveny, R.S., and R.C. Balling Jr. 1998. Weekly cycles of air pollutants, precipitation and tropical cyclones in the coastal NW Atlantic region. *Nature* 394: 561–563.

Chameides, W.L., F. Fehsenfeld, M.O. Rodgers, C. Cardelino, J. Martinez, D. Parrish, W. Lonneman, D.R. Lawson, R.A. Rasmussen, P. Zimmerman, J. Greenberg, P. Middleton, and T. Wang.

1992. Ozone precursor relationships in the ambient atmosphere. *Journal of Geophysical Research* 97: 6037–6055.

Chock, D.L., T.Y. Chang, S.L. Winkler, and B.I. Nance. 1999. The impact of an 8h ozone air quality standard on ROG and NO_x controls in Southern California. *Atmospheric Environment* 33: 2471–2485.

Cleveland, C.C., A.R. Townsend, D.S. Schimel, H. Fisher, R.W. Howarth, L.O. Hedin, S.S. Perakis, E.F. Latty, J.C. von Fischer, A. Elseroad, and M.F. Wasson. 1999. Global patterns of terrestrial biological nitrogen (N_2) fixation in natural systems. *Global Biogeochemical Cycles* 13: 623–645.

Cleveland, C.J., R. Costanza, C.A.S. Hall, and R. Kaufmann. 1984. Energy and the U.S. economy: A biophysical perspective. *Science* 225: 890–897.

Collins, J.P., A. Kinzig, N.B. Grimm, W.F. Fagan, D. Hope, J. Wu, and E.T. Borer. 2000. A new urban ecology: Modeling human communities as integral parts of ecosystems poses special problems for the development and testing of ecological theory. *American Scientist* 88: 416–425.

Cooper, S.R., S.K. McGlothlin, M. Madritch, and D.L. Jones. 2004. Paleoecological evidence of human impacts on the Neuse and Pamlico estuaries of North Carolina, USA. *Estuaries* 27: 617–633.

Costanza, R. 2006. Nature: Ecosystems without commodifying them. *Nature* 443: 749.

Costanza, R., and H.E. Daly. 1992. Natural capital and sustainable development. *Conservation Biology* 6: 37–46.

Costanza, R., R. d'Arge, R. deGroot, S. Farber, M. Grasso, B. Hannon, K. Limburg, S. Naeem, R.V. O'Neill, J. Paruelo, R.G. Raskin, P. Sutton, and M. van den Belt. 1997. The value of the world's ecosystem services and natural capital. *Nature* 387: 253–260.

Currie, D.J. 1991. Energy and large-scale patterns of animal- and plant species richness. *The American Naturalist* 137: 27–49.

D'Amato, G., G. Liccardi, and G. Frenguelli. 2007. Thunderstorm-asthma and pollen allergy. *Allergy* 62: 11–16.

Davis, R.E., P.C. Knappenberger, P.J. Michaels, and W.M. Novicoff. 2003. Changing heat-related mortality in the United States. *Environmental Health Perspectives* 111: 1712–1718.

Davis, R.E., P.C. Knappenberger, P.J. Michaels, and W.M. Novicoff. 2004. Seasonality of climate–human mortality relationships in US cities and impacts of climate change. *Climate Research* 26: 61–76.

de Gouw, J.A., A.M. Middlebrook, C. Warneke, P.D. Goldan, W.C. Kuster, J.M. Roberts, F.C. Fehsenfeld, D.R. Worsnop, M.R. Canagaratna, A.A.P. Pszenny, W.C. Keene, M. Marchewka, S.B. Bertman, and T.S. Bates. 2005. Budget of organic carbon in a polluted atmosphere: Results from the New England Air Quality Study in 2002. *Journal of Geophysical Research* 110: D16305.

DeLucia, E.H., D.J. Moore, and R.J. Norby. 2005. Contrasting responses of forest ecosystems to rising atmospheric CO_2: Implications for the global C cycle. Global Biogeochem. *Cycles* 19: GB3006.

DeWald, S., S. Josiah, and B. Erdkamp. 2005. Heating with wood: Producing, harvesting and processing firewood. University of Nebraska–Lincoln Extension Publication G1554.

Dindorf, T., U. Kuhn, L. Ganzeveld, G. Schebeske, P. Ciccioli, C. Holzke, R. Köble, G. Seufert, and J. Kesselmeier. 2006. Significant light and temperature dependent monoterpene emissions from European beech (*Fagus sylvatica* L.) and their potential impact on the European volatile organic compound budget. *Journal of Geophysical Research* 111: D16305.

Dixon, P.G., and T.L. Mote. 2003. Patterns and causes of Atlanta's urban heat island–initiated precipitation. *Journal of Applied Meteorology* 42: 1273–1284.

Dockins, C., K. Maguire, N. Simon, and M. Sullivan. 2004. Value of statistical life analysis and environmental policy: A white paper. Environmental Ecolonomics Report #0483.

Donnelly, R., and J.M. Marzluff. 2006. Relative importance of habitat quantity, structure, and spatial pattern to birds in urbanizing environments. *Urban Ecosystems* 9: 99–117.

Downey, L. 2007. US metropolitan-area variation in environmental inequality outcomes. *Urban Studies* 44: 953–977.

Druffel, E.R.M., S. Griffin, T.P. Guilderson, M. Kashgarian, J. Southon, and D.P. Schrag. 2001. Changes of subtropical north Pacific radiocarbon and correlation with climate variability. *Radiocarbon* 43: 15–25.

Dukes, J.S. 2003. Burning buried sunshine: Human consumption of ancient solar energy. *Climatic Change* 61: 31–44.

Dukes, J.S., and H.A. Mooney. 2004. Disruption of ecosystem processes in western North America by invasive species. *Revista Chilena de Historia Natural* 77: 411–437.

Dunn, P.O., and D.W. Winkler. 1999. Climate change has affected the breeding date of tree swallows throughout North America. *Proceedings of the Royal Society of London* B 266: 2487–2490.

Ebdon, J.S., A.M. Petrovic, and R.A. White. 1999. Interaction of nitrogen, phosphorus, and potassium on evapotranspiration rate and growth of Kentucky bluegrass. *Crop Science* 39: 209–218.

Ellis, C.D., S. Lee, and B. Kweon. 2006. Retail land use, neighborhood satisfaction and the urban forest: An investigation into the moderating and mediating effects of trees and shrubs. *Landscape and Urban Planning* 74: 70–78.

Ezzati, M., A.B. Friedman, S.C. Kulkarni, and C.J.L. Murray. 2008. The reversal of fortunes: Trends in county mortality and cross-county mortality disparities in the United States. *PLoS Medicine* 5(4): e66.

Fagan, W.F., E. Meir, S.S. Carroll, and J. Wu. 2001. The ecology of urban landscapes: Modeling housing starts as a density-dependent colonization process. *Landscape Ecology* 16: 33–39.

Falkowski, P., R.J. Scholes, E. Boyle, J. Canadell, D. Canfield, J. Elser, N. Gruber, K. Hibbard, P. Hgberg, S. Linder, F.T. Mackenzie, B. Moore III, T. Pedersen, Y. Rosenthal, S. Seitzinger, V. Smetacek, and W. Steffen. 2000. The global carbon cycle: A test of our knowledge of Earth as a system. *Science* 290: 291–296.

Farber, S., R. Costanza, D.L. Childers, J. Erickson, K. Gross, M. Grove, C.S. Hopkinson, J. Kahn, S. Pincetl, A. Troy, P. Warren, and M. Wilson. 2006. Linking Ecology and Economics for Ecosystem Management. *BioScience* 56: 117–129.

Farber, S.C., R. Costanza, and M.A. Wilson. 2002. Economic and ecological concepts for valuing ecosystem services. *Ecological Economics* 41: 375–392.

Felson, A.J., and S.T.A. Pickett. 2005. Designed experiments: New approaches to studying urban ecosystems. *Frontiers in Ecology and the Environment* 3: 549–556.

Flores, A., S.T.A. Pickett, W.C. Zipperer, R.V. Pouyat, and R. Pirani. 1998. Adopting a modern ecological view of the metropolitan landscape: The case of a greenspace system for the New York City region. *Landscape and Urban Planning* 39: 295–308.

Food and Agriculture Organization (FAO) of the United Nations. 2003. Review of world water resources by country. Report #23, Rome.

Forbes, S.A. 1880. The food of birds. *Bulletin of the Illinois State Laboratory of Natural History* 1(3): 80–148.

Franco, G., and A.H. Sanstad. 2006. Climate change and electricity demand in California. California Climate Change Center Report CEC-500-2005-201-SF.

Freer-Smith, P.H., A.A. El-Khatib, and G. Taylor. 2004. Capture of particulate pollution by trees: A comparison of species typical of semi-arid areas (*Ficus nitida* and *Eucalyptus globulus*) with European and North American species. *Water, Air, and Soil Pollution* 155: 173–187.

Friedman, M.S., K.E. Powell, L. Hutwagner, L.M. Graham, and W.G. Teague. 2001. Impact of changes in transportation and commuting behaviors during the 1996 summer Olympic Games in Atlanta on air quality and childhood asthma. *Journal of the American Medical Association* 285: 897–905.

Fuller, R.A., K.N. Irvine, P. Devine-Wright, P.H. Warren, and K.J. Gaston. 2007. Psychological benefits of greenspace increase with biodiversity. *Biology Letters* 3: 390–394.

Galan, I., A. Tobias, J.R. Banegas, and E. Aranguez. 2003. Short-term effects of air pollution on daily asthma emergency room admissions. *European Respiratory Journal* 22: 802–808.

Galloway, J.N., J.D. Aber, J.W. Erisman, S.P. Seitzinger, R.W. Howarth, E.B. Cowling, and J. Cosby. 2003. The nitrogen cascade. *Bioscience* 53: 341–356.

Galloway, J.N., and E.B. Cowling. 2002. Reactive nitrogen and the world: 200 years of change. *Ambio* 31: 64–71.

Gates, D.M. 1965. Energy, plants, and ecology. *Ecology* 46: 1–13.

Gates, D.M., H.J. Keegan, J.C. Schleter, and V.R. Weidner. 1965. Spectral Properties of Plants. *Applied Optics* 4: 11–20.

Georgi, N.J., and K. Zafiriadis. 2006. The impact of park trees on microclimate in urban areas. *Urban Ecosystems* 9: 195–209.

Geron, C., P. Harley, and A. Guenther. 2001. Isoprene emission capacity for US tree species. *Atmospheric Environment* 35: 3341–3352.

Gidlöf-Gunnarsson, A., and E. Öhrström. 2007. Noise and well-being in urban residential environments: The potential role of perceived availability to nearby green areas. *Landscape and Urban Planning* 83: 115–126.

Gillani N.V., S. Kohli, and W.E. Wilson. 1981. Gas-to particle conversion of sulfur in power plant plumes: I. Parameterization of the conversion rate for moderately polluted ambient conditions. *Atmospheric Environment* 15: 2293–2313.

Gillani, N.V., and J.E. Pleim. 1996. Sub-grid-scale features of anthropogenic emissions of NOx and VOC in the context of regional Eulerian models. *Atmospheric Environment* 30: 2043–2059.

Gillani N.V., and Y. Wu. April 2003. Exploration of uncertainty in the simulation of power plant chemistry. Final Report of University of Alabama in Huntsville to EPA STAR Grant GR826239-01-0.

Givoni, B. 1991. Impact of planted areas on urban environmental quality: A review. *Atmospheric Environment* 25B: 289–299.

Graham, R.L., L.L. Wright, and A.F. Turhollow. 1992. The potential for short-rotation woody crops to reduce US CO2 emissions. *Climatic Change* 22: 223–238.

Groffman, P.M., N.L. Law, K.T. Belt, L.E. Band, and G.T. Fisher. 2004. Nitrogen fluxes and retention in urban watershed ecosystems. *Ecosystems* 7: 393–403.

Grove, J.M., and W.R. Burch Jr. 1997. A social ecology approach and applications of urban ecosystem and landscape analyses: A case study of Baltimore, Maryland. *Urban Ecosystems* 1: 259–275.

Grove, J.M., A.R. Troy, J.P.M. O'Neil-Dunne, W.R. Burch Jr., M.L. Cadenasso, and S.T.A. Pickett. 2006. Characterization of households and its implications for the vegetation of urban ecosystems. *Ecosystems* 9: 578–597.

Guenther, A., T. Karl, P. Harley, C. Wiedinmyer, P. Palmer, and C. Geron. 2006. Estimates of global terrestrial isoprene emissions using MEGAN (Model of Emissions of Gases and Aerosols from Nature). *Atmospheric Chemistry and Physics* 6: 3181–3210.

Guenther, A., P. Zimmerman, and M. Wildermuth. 1994. Natural volatile organic compound emission rate estimates for U.S. woodland landscapes. *Atmospheric Environment* 28: 1197–1210.

Guillebeau, P., N. Hinkle, and P. Roberts, eds. 2006. Summary of losses from insect damage and cost of control in Georgia 2003. The University of Georgia, College of Agriculture and Environmental Sciences, Department of Entomology Special Committee on Insect Surveys and Losses, Publication #106.

Haire, S.L., C.E. Bockb, B.S. Cadea, and B.C. Bennett. 2000. The role of landscape and habitat characteristics in limiting abundance of grassland nesting songbirds in an urban open space. *Landscape and Urban Planning* 48: 65–82.

Hall, D.C. 1998. Albedo and vegetation demand-side management options for warm climates. *Ecological Economics* 24: 31–45.

Hanemann, W.M. 1991. Willingness to pay and willingness to accept: How much can they differ? *The American Economic Review* 81: 635–647.

Hanemann, W.M. 1994. Valuing the environment through contingent valuation. *The Journal of Economic Perspectives* 8: 19–43.

Hanemann, W.H. 2006. The economic conception of water. In *Water Crisis: Myth or Reality*? Eds. P.P. Rogers, M.R. Llamas, and L. Martinez-Cortina. Taylor & Francis.

Hansen, A.J., R.L. Knight, J.M. Marzluff, S. Powell, K. Brown, P.H. Gude, and K. Jones. 2005. Effects of exurban development on biodiversity: Patterns, mechanisms, and research needs. *Ecological Applications* 15: 1893–1905.

Harlan, S.L., A.J. Brazel, L. Prashad, W.L. Stefanov, and L. Larsen. 2006. Neighborhood microclimates and vulnerability to heat stress. *Social Science & Medicine* 63: 2847–2863.

Harley, P., A. Guenther, and P. Zimmerman. 1997. Environmental controls over isoprene emission in deciduous oak canopies. *Tree Physiology* 17: 705–714.

Harrison, G.W. 1992. Valuing public goods with the contingent valuation method: A critique of Kahneman and Knetsch. *Journal of Environmental Economics and Management* 23: 248–257.

Hasenauer, H. 1997. Dimensional relationships of open-grown trees in Austria. *Forest Ecology and Management* 96: 197–206.

Hashimoto, M., T. Fukuda, T. Shimizu, S. Watanabe, S. Watanuki, Y. Eto, and M. Urashima. 2004. Influence of climate factors on emergency visits for childhood asthma attack. *Pediatrics International* 46: 48–52.

Heilman, J.L., and R.W. Gesch. 1991. Effects of turfgrass evaporation on external temperatures of buildings. *Theoretical and Applied Climatology* 43: 185–194.

Heynen, N., H.A. Perkins, and P. Roy. 2006. The political ecology of uneven urban green space: The impact of political economy on race and ethnicity in producing environmental inequality in Milwaukee. *Urban Affairs Review* 42: 3–25.

Hicke, J.A., G.P. Asner, J.T. Randerson, C. Tucker, S. Los, R. Birdsey, J.C. Jenkins, and C. Field. 2002. Trends in North American net primary productivity derived from satellite observations, 1982–1998. *Global Biogeochemical Cycles* 16: 1018.

Hofmockel, K.S., and W.H. Schlesinger. 2007. Carbon dioxide effects on heterotrophic dinitrogen fixation in a temperate pine forest. *Soil Science Society of America Journal* 71: 140–144.

Hoover, M.D. 1944. Effect of removal of forest vegetation upon water-yields. *Transactions American Geophysical Union*, pp. 969–977.

Hope, D., C. Gries, W. Zhu, W.F. Fagan, C.L. Redman, N.B. Grimm, A.L. Nelson, C. Martin, and A. Kinzig. 2003. Socioeconomics drive urban plant diversity. *Proceedings of the National Academy of Sciences* 100: 8788–8792.

Hoppe, R.A., P. Korb, E.J. O'Donoghue, and D.E. Banker. 2007. Structure and finances of U.S. farms: Family farm report, 2007 edition. United States Department of Agriculture, Economic Infomation Bulletin Number 24.

Horowitz, J.K., and K.E. McConnell. 2002. A Review of WTA/WTP Studies. *Journal of Environmental Economics and Management* 44: 426–447.

Houck, J.E., P.E. Tiegs, R.C. McCrillis, C. Keithley, and J. Crouch. 1998. Air emissions from residential heating: The wood heating option put into environmental perspective. In 8th AWMA Conference, December 8–10, The Emission Inventory: Living in a Global Environment, New Orleans, pp. 373–384.

Howard, L. 1833. The climate of London, deduced from meteorological observations, made in the metropolis, and at various places around it. Harvey and Darton (London).

Huang, Y.J., H. Akbari, H. Taha, and A. Rosenfeld. 1987. The potential of vegetation in reducing summer cooling loads in residential buildings. *Climate and Applied Meteorology* 26: 1103–1116.

Jo, H.-K., and E.G. McPherson. 1995. Carbon storage and flux in urban residential greenspace. *Journal of Environmental Management* 45: 109–133.

Kalkstein, L.S., and R.E. Davis. 1989. Weather and human mortality: An evaluation of demographic and interregional responses in the United States. *Annals of the Association of American Geographers* 79: 44–64.

Kalkstein, L.S., and J.S. Greene. 1997. An evaluation of climate/mortality relationships in large U.S. cities and the possible impacts of a climate change. *Environmental Health Perspectives* 105: 84–93.

Karl, T.R., H.F. Diaz, and G. Kukla. 1988: Urbanization: Its detection and effect in the United States climate record. *Journal of Climate* 1: 1099–1123.

Karnosky, D.F., J.M. Skelly, K.E. Percy, A.H. Chappelka. 2007. Perspectives regarding 50 years of research on effects of tropospheric ozone air pollution on US forests. *Environmental Pollution* 147: 489–506.

Kaushal, S.S., P.M. Groffman, L.E. Band, C.A. Shields, R.P. Morgan, M.A. Palmer, K.T. Belt, C.M. Swan, S.E.G. Findlay, and G.T. Fisher. 2008. Interaction between urbanization and climate variability amplifies watershed nitrate export in Maryland. *Environmental Sciences of Technology* 42: 5872–5878.

Kaushal, S.S., P.M. Groffman, G.E. Likens, K.T. Belt, W.P. Stack, V.R. Kelly, L.E. Band, and G.T. Fisher. 2005. Increased salinization of fresh water in the northeastern United States. *Proceedings of the National Academy of Sciences* 102: 13517–13520.

Kaye, J.P., R.L. McCulley, and I.C. Burke. 2005. Carbon fluxes, nitrogen cycling, and soil microbial communities in adjacent urban, native and agricultural ecosystems. *Global Change Biology* 11: 575–587.

Keeling, C.D., S.C. Piper, R.B. Bacastow, M. Wahlen, T.P. Whorf, M. Heimann, and H.A. Meijer. 2005. Atmospheric CO_2 and $^{13}CO_2$ exchange with the terrestrial biosphere and oceans from 1978 to 2000: Observations and carbon cycle implications, pp. 83–113. In *A History of Atmospheric CO_2 and Its Effects on Plants, Animals, and Ecosystems*. Eds. J.R. Ehleringer, T.E. Cerling, and M.D. Dearing. Springer Verlag.

Kiehl, J.T., and K.E. Trenberth. 1997. Earth's annual global mean energy budget. *Bulletin of the American Meteorological Society* 78: 197–208.

Kinney, P.L., K. Knowlton, and C. Hogrefe. 2005. Ozone: Kinney et al. Respond. *Environmental Health Perspectives* 113: A87.

Kjelgren, R., and T. Montague. 1998. Urban tree transpiration over turf and asphalt surfaces. *Atmospheric Environment* 32: 35–41.

Knight, A.J. 2008. "Bats, snakes and spiders, Oh my!" How aesthetic and negativistic attitudes, and other concepts predict support for species protection. *Journal of Environmental Psychology* 28: 94–103.

Knowlton, K., J.E. Rosenthal, C. Hogrefe, B. Lynn, S. Gaffin, R. Goldberg, C. Rosenzweig, K. Civerolo, J.-Y. Ku, and P.L. Kinney. 2004. Assessing ozone-related health impacts under a changing climate. *Environmental Health Perspectives* 112: 1557–1563.

Krajicek, J.E., K.A. Brinkman, and S.F. Gingrich. 1961. Crown competition: A measure of density. *Forest Science* 7: 35–42.

Kuo, F.E., and W.C. Sullivan. 2001a. Environment and crime in the inner city: Does vegetation reduce crime? *Environment and Behavior* 33: 343–367.

Kuo, F.E., and W.C. Sullivan. 2001b. Aggression and violence in the inner city: Effects of environment via mental fatigue. *Environment and Behavior* 33: 543–571.

Kuttler, W., and A. Strassburger. 1999. Air quality measurements in urban green areas — a case study. *Atmospheric Environment* 33: 4101–4108.

Lal, R. 2003. Global potential of soil carbon sequestration to mitigate the greenhouse effect. *Critical Reviews in Plant Sciences* 22: 151–184.

Lerdau, M., A. Guenther, and R. Monson. 1997. Plant production and emission of volatile organic compounds. *BioScience* 47: 373–383.

Levin, I., and V. Hesshaimer. 2000. Radiocarbon: A unique tracer of global carbon cycle dynamics. *Radiocarbon* 42: 69–80.

Likens, G.E., D.C. Buso, and T.J. Butler. 2005. Long-term relationships between SO_2 and NO_x emissions and SO_4^{2-} and NO_3^- concentration in bulk deposition at the Hubbard Brook Experimental Forest, NH. *Journal of Environmental Monitoring* 7: 964–968.

Lin, S., S.-A. Hwang, C. Pantea, C. Kielb, and E. Fitzgerald. 2004. Childhood asthma hospitalizations and ambient air sulfur dioxide concentrations in Bronx County, New York. *Archives of Environmental Health* 59: 266–275.

Liu, G.C., J.S. Wilson, R. Qi, and J. Ying. 2007. Green neighborhoods, food retail and childhood overweight: Differences by population density. *American Journal of Health Promotion* 21: 317–325.

Lovett, G.M. 1994. Atmospheric deposition of nutrients and pollutants in North America: An ecological perspective. *Ecological Applications* 4: 629–650.

Lu, J., G. Sun, S.G. McNulty, and D.M. Amatya. 2003. Modeling actual evapotranspiration from forested watersheds across the southeastern United States. *Journal of American Water Resources Association* 39: 887–896.

Lubowski, R.N., M. Vesterby, S. Bucholtz, A. Baez, and M.J. Roberts. 2006. Major uses of land in the United States, 2002. USDA Economic Research Service Bulletin 14.

Luria, M., R.E. Imhoff, R.J. Valente, and R.L. Tanner. 2003. Ozone yields and production efficiencies in a large power plant plume. *Atmospheric Environment* 37: 3593–3603.

Maantay, J. 2007. Asthma and air pollution in the Bronx: Methodological and data considerations in using GIS for environmental justice and health research. *Health and Place* 13: 32–56.

MacDonald, J., R. Hoppe, and D. Banker. 2006. Growing farm size and the distribution of farm payments. USDA Economic Brief #6.

Marland, G., R. Andres, and T. Boden. 2008. Global CO2 emissions from fossil-fuel burning, cement manufacture, and gas flaring: 1751–2005. Carbon Dioxide Information Analysis Center, Oak Ridge National Laboratory.

Marvier, M., J. Grant, and P. Kareiva. 2006. Nature: Poorest may see it as their economic rival. *Nature* 443: 749–750.

Maurellis, A., and J. Tennyson. May 2003. The climatic effects of water vapour. *Physics World*.

McCarthy, M.A., C.J. Thompson, and N.S.G. Williams. 2006. Logic for designing nature reserves for multiple species. *American Naturalist* 167: 717–727.

McCauley, D.J. 2006. Selling out on nature. *Nature* 443: 27–28.

McDonald, A.G., W.J. Bealey, D. Fowler, U. Dragosits, U. Skiba, R.I. Smith, R.G. Donovan, H.E. Brett, C.N. Hewitt, and E. Nemitz. 2007. Quantifying the effect of urban tree planting on concentrations and depositions of PM_{10} in two UK conurbations. *Atmospheric Environment* 41: 8455–8467.

McGeehin, M.A., and M. Mirabelli. 2001. The potential impacts of climate variability and change on temperature-related morbidity and mortality in the United States. *Environmental Health Perspectives* 109: 185–189.

McPherson, E.G. 2001. Sacramento's parking lot shading ordinance: environmental and economic costs of compliance. *Landscape and Urban Planning* 57: 105–123.

McPherson, E.G, K.I. Scott, and J.R. Simpson. 1998. Estimating cost effectiveness of residential yard trees for improving air quality in Sacramento, California, using existing models. *Atmospheric Environment* 32: 75–84.

McPherson, E.G., J.R. Simpson, and M. Livingston. 1989. Effects of three landscape treatments on residential energy and water use in Tucson, Arizona. *Energy and Buildings* 13: 127–138.

McPherson, E.G, J.R. Simpson, P.J. Peper, S.L. Gardner, K.E. Vargas, S.E. Maco, and Q. Xiao. 2006. Piedmont Community tree guide: Benefits, costs, and strategic planting. USDA Forest Service, State and Private Forestry, Urban and Community Forestry Program.

McPherson, E.G, J.R. Simpson, P.J. Peper, S.E. Maco, and Q. Xiao. 2005. Municipal forest benefits and costs in five US cities. *Journal of Forestry* 103: 411–416.

McPherson, E.G, J.R. Simpson, P.J. Peper, and Q. Xiao. 1999. Tree guidelines for San Joaquin Valley communities. USDA Forest Service, Pacific Southwest Research Station.

Mennis, J. 2006. Socioeconomic-vegetation relationships in urban, residential land: The case of Denver, Colorado. *Photogrammetric Engineering & Remote Sensing* 72: 911–921.

Michael, S.E., R.B. Hull, and D.L. Zahm. 2001. Environmental factors influencing auto burglary: A case study. *Environment and Behavior* 33: 368–388.

Minino, A.M., M.P. Heron, S.L. Murphy, and K.D. Kochanek. 2007. Deaths: Final data for 2004. National Vital Statistics Reports, Vol. 55, No. 19. Hyattsville, MD: National Center for Health Statistics.

Mitchell, G., and D. Dorling. 2003. An environmental justice analysis of British air quality. *Environment and Planning* A 35: 909–929.

Mitchell, R., and F. Popham. 2007. Greenspace, urbanity and health: Relationships in England. *Journal of Epidemiology and Community Health* 61: 681–683.

Moore, K., R. Neugebauer, F. Lurmann, J. Hall, V. Brajer, S. Alcorn, and I. Tager. 2008. Ambient ozone concentrations cause increased hospitalizations for asthma in children: An 18-year study in Southern California. *Environmental Health Perspectives* 116: 1063–1070.

National Agricultural Statistics Service, U.S. Department of Agriculture, www.nass.usda.gov/QuickStats.

National Research Council, Committee on Reducing Stormwater Contributions to Water Pollution, 2008. *Urban Stormwater Management in the United States*. National Academy Press.

Neil, K., and J. Wu. 2006. Effects of urbanization on plant flowering phenology: A review. *Urban Ecosystems* 9: 243–257.

Nobel, P.S. 1991. *Physicochemical and Environmental Plant Physiology*. Academic Press.

Nowak, D.J. 1994. Atmospheric carbon dioxide reduction by Chicago's urban forest. In *Chicago's Urban Forest Ecosystem: Results of the Chicago Urban Forest Climate Project*. Eds. E.G. McPherson, D.J. Nowak, and R.A. Rowntree. USDA Forest Service General Technical Report NE-186, Radnor, PA, pp. 83–94.

Nowak, D.J., and D.E. Crane. 2002. Carbon storage and sequestration by urban trees in the USA. *Environmental Pollution* 116: 381–389.

Nowak, D.J., D.E. Crane, and J.F. Dwyer. 2002a. Compensatory value of urban trees in the U.S. *Journal of Arboriculture* 28: 194–199.

Nowak, D.J., J.C. Stevens, S.M. Sisinni, and C.J. Luley. 2002b. Effects of urban tree management and species selection on atmospheric carbon dioxide. *Journal of Arboriculture* 28: 113–122.

Oke, T.R. 1973. City size and the urban heat island. *Atmospheric Environment* 7: 769–779.

Oke, T.R. 1982. The energetic basis of the urban heat island. *Quarterly Journal of the Royal Meteorological Society* 108: 1–24.

Olszyna, K.J., M. Luria, and J.F. Meagher. 1997. The correlation of temperature and rural ozone levels in southeastern U.S.A. *Atmospheric Environment* 31: 3011–3022.

Orendovici, T., J.M. Skelly, J.A. Ferdinand, J.E. Savage, M.-J. Sanz, and G.C. Smith. 2003. Response of native plants of northeastern United States and southern Spain to ozone exposures: Determining exposure/response relationships. *Environmental Pollution* 125: 31–40.

Ostfeld, R.S., and F. Keesing. 2000. Biodiversity and disease risk: The case of Lyme disease. *Conservation Biology* 14: 722–728.

Pan, Y., R. Birdsey, J. Hom, K. McCullough, and K. Clark. 2006. Improved estimates of net primary productivity from MODIS satellite data at regional and local scales. *Ecological Applications* 16: 125–132.

Parker, D., J. Sonne, and J. Sherwin. 1997. Demonstration of cooling savings of light colored roof surfacing in Florida commercial buildings: Retail strip mall. Final Report to the Florida Solar Energy Center, University of Central Florida.

Parker, D., J. Sonne, and J. Sherwin. 2003. Flexible roofing facility: 2002 summer test results. Florida Solar Energy Center Publication FSEC-CR-1411-03.

Parker, D.E. 2006. A demonstration that large-scale warming is not urban. *Journal of Climate* 19: 2882–2895.

Parker, D.S., S.F. Barkaszi, Jr., S. Chandra, and D.J. Beal. 1995. Measured cooling energy savings from reflective roofing systems in Florida: Field and laboratory research results. *Proceedings of the Thermal Performance of the Exterior Envelopes of Buildings VI*, December 4–8, Clearwater, FL (cited in Akbari et al. 2001).

Parker, J.H. 1981. Use of landscaping for energy conservation. Department of Physical Sciences, Florida International University, Miami, FL (cited in Akbari 2002).

Parris, K.M., and D.L. Hazell. 2005. Biotic effects of climate change in urban environments: The case of the grey-headed flying-fox (*Pteropus poliocephalus*) in Melbourne, Australia. *Biological Conservation* 124: 267–276.

Parris, K.M., and A. Schneider 2008. Impacts of traffic noise and traffic volume on birds of roadside habitats. *Ecology and Society* 14: 29.

Parris, K.M., M. Velik-Lord, and J.M.A. North. 2009. Frogs call at a higher pitch in traffic noise. *Ecology and Society* 14: 25.

Parrish, D.D., J.S. Holloway, M. Trainer, P.C. Murphy, G.L. Forbes, and F.C. Fehsenfeld. 1993. Export of North American ozone pollution to the North Atlantic ocean. *Science* 259: 1436–1439.

Peel, J.L., P.E. Tolbert, M. Klein, K.B. Metzger, W.D. Flanders, K. Todd, J.A. Mulholland, P.B. Ryan, and H. Frumkin. 2005. Ambient air pollution and respiratory emergency department visits. *Epidemiology* 16: 164–174.

Perkins, H.A., N. Heynen, and J. Wilson. 2004. Inequitable access to urban reforestation: The impact of urban political economy on housing tenure and urban forests. *Cities* 21: 291–299.

Peters, R.H. 1983. *Ecological Implications of Body Size.* Cambridge University Press.

Peterson, T.C. 2003. Assessment of urban versus rural in situ surface temperatures in the contiguous United States: No difference found. *Journal of Climate* 16: 2941–2959.

Peterson, T.C., and T.W. Owen. 2005. Urban heat island assessment: Metadata are important. *Journal of Climate* 18: 2637–2646.

Petit, J.R., J. Jouzel, D. Raynaud, N.I. Barkov, J.-M. Barnola, I. Basile, M. Bender, J. Chappellaz, M. Davis, G. Delaygue, M. Delmotte, V.M. Kotlyakov, M. Legrand, V.Y. Lipenkov, C. Lorius, L. Pépin, C. Ritz, E. Saltzman, and M. Stievenard. 1999. Climate and atmospheric history of the past 420,000 years from the Vostok ice core, Antarctica. *Nature* 399: 429–436.

Pickett, S.T.A., and M.L. Cadenasso. 2006. Advancing urban ecological studies: Frameworks, concepts, and results from the Baltimore Ecosystem Study. *Austral Ecology* 31: 114–125.

Pickett, S.T.A., M.L. Cadenasso, J.M. Grove, C.H. Nilon, R.V. Pouyat, W.C. Zipperer, and R. Costanza. 2001. Urban ecological systems: Linking terrestrial ecological, physical, and socioeconomic components of metropolitan areas. *Annual Review of Ecology and Systematics* 32: 127–57.

Pleijel, H., H. Danielsson, L. Emberson, M.R. Ashmore, and G. Mills. 2007. Ozone risk assessment for agricultural crops in Europe: Further development of stomatal flux and flux — response relationships for European wheat and potato. *Atmospheric Environment* 41: 3022–3040.

Pomerantz, M., H. Akbari, P. Berdahl, S.J. Konopacki, H. Taha, and A.H. Rosenfeld. 1999. Reflective surfaces for cooler buildings and cities. *Philosophical Magazine* B 79: 1457–1476.

Potter, C., S. Klooster, R. Nemani, V. Genovese, S. Hiatt, M. Fladeland, and P. Gross. 2006. Estimating carbon budgets for U.S. ecosystems. *Eos* 87: 85–96.

Pouyat, R.V., I.D. Yesilonis, and D.J. Nowak. 2006. Carbon storage by urban soils in the United States. *Journal of Environmental Quality* 35: 1566–1575.

Pozzi, F., and C. Small. 2001. Exploratory analysis of suburban land cover and population density in the U.S.A. Proceedings of the IEEE Joint Workshop on Remote Sensing and Data Fusion over Urban Areas, November 8–9, Rome, Italy, pp. 250–254.

Pozzi, F., and C. Small. 2005. Analysis of urban land cover and population density in the United States. *Photogrammetric Engineering & Remote Sensing* 71: 719–726.

Presterl, T., S. Groh, M. Landbeck, G. Seitz, W. Schmidt, and H.H. Geiger. 2002. Nitrogen uptake and utilization efficiency of European maize hybrids developed under conditions of low and high nitrogen input. *Plant Breeding* 121: 480–486.

Price, L., C. Marnay, J. Sathaye, S. Murtishaw, D. Fisher, A. Phadke, and G. Franco 2002. The California Climate Action Registry: Development of methodologies for calculating greenhouse gas emissions from electricity generation. Proceedings of the ACEEE Summer Study on Energy Efficiency in Buildings, Pacific Grove, CA.

Qian, Y., and R.F. Follett. 2002. Assessing soil carbon sequestration in turfgrass systems using long-term soil testing data. *Agronomy Journal* 94: 930–935.

Raven, P.H., R.F. Evert, and S.E. Eichhorn. 1986. *The Biology of Plants*. Worth Publishers.

Rees, W.E. 1992. Ecological footprints and appropriated carrying capacity: What urban economics leaves out. *Environment and Urbanization* 4: 121–130.

Rees, W.E. 1998. How should a parasite value its host? *Ecological Economics* 25: 49–52.

Reid, W.V. 2006. Nature: The many benefits of ecosystem services. *Nature* 443: 749.

Riley, S.P.D., G.T. Busteed, L.B. Kats, T.L. Vandergon, L.F.S. Lee, R.G. Dagit, J.L. Kerby, R.N. Fisher, and R.M. Sauvajot. 2005. Effects of urbanization on the distribution and abundance of amphibians and invasive species in Southern California streams. *Conservation Biology* 19: 1894–1907.

Roetzer, T., M. Wittenzeller, H. Haeckel, and J. Nekovar. 2000. Phenology in central Europe: Differences and trends of spring phenophases in urban and rural areas. *International Journal of Biometeorology* 44: 60–66.

Rojstaczer, S., S.M. Sterling, and N.J. Moore. 2001. Human appropriation of photosynthesis products. *Science* 294: 2549–2552.

Ryerson, T.B., M. Trainer, J.S. Holloway, D.D. Parrish, L.G. Huey, D.T. Sueper, G.J. Frost, S.G. Donnelly, S. Schauffler, E.L. Atlas, W.C. Kuster, P.D. Goldan, G. Hubler, J.F. Meagher, and F.C. Fehsenfeld. 2001. Observations of ozone formation in power plant plumes and implications for ozone control strategies. *Science* 292: 719–723.

Scharenbroch, B.C., J.E. Lloyd, and J.L. Johnson-Maynard. 2005. Distinguishing urban soils with physical, chemical, and biological properties. *Pedobiologia* 49: 283–296.

Schlesinger, W.H. 1997. *Biogeochemistry: An Analysis of Global Change*. Academic Press.

Schlesinger, W.H., and W.A. Reiners. 1974. Deposition of water and cations on artificial foliar collectors in fir krummholz of New England mountains. *Ecology* 55: 378–385.

Schmer, M.R., K.P. Vogel, R.B. Mitchell, and R.K. Perrin. 2008. Net energy of cellulosic ethanol from switchgrass. *Proceedings of the National Academy of Sciences* 105: 464–469.

Schor, J.B. 2005. Prices and quantities: Unsustainable consumption and the global economy. *Ecological Economics* 55: 309–320.

Schroeder, H.W., and L.M. Anderson. 1984. Perception of personal safety in urban recreation sites. *Journal of Leisure Research* 16: 178–194.

Schwartz, M.D., R. Ahas, and A. Aasa. 2006. Onset of spring starting earlier across the Northern Hemisphere. *Global Change Biology* 12: 343–351.

Seinfeld, J.H., and S.N. Pandis. 2006. *Atmospheric Chemistry and Physics: From Air Pollution to Climate Change, Second Edition.* John Wiley & Sons.

Seneviratne, S.I., D. Luthi, M. Litschi, and C. Schar. 2006. Land–atmosphere coupling and climate change in Europe. *Nature* 443: 205–209.

Sexton, J.O., D.L. Urban, and C. Song. In review. Signature extension of the 2001 National Landcover Dataset (NLCD 2001) Tree Canopy Layer for Piedmont North Carolina. Remote Sensing of Environment.

Shackleton, N.J. 2000. The 100,000-year ice-age cycle identified and found to lag temperature, carbon dioxide, and orbital eccentricity. *Science* 289: 1897–1902.

Shafik, N., and S. Bandyopadhyay. 1992. Economic growth and environmental quality: Time-series and cross-country evidence. *World Development Report* WPS-904.

Shimano, K. 1997. Analysis of the relationship between DBH and crown projection area using a new model. *Journal of Forest Research* 2: 237–242.

Sillman, S. 1999. The relation between ozone, NO_x and hydrocarbons in urban and polluted rural environments. *Atmospheric Environment* 33: 1821–1845.

Simpson, J.R. 1998. Urban forest impacts on regional cooling and heating energy use: Sacramento county case study. *Journal of Arboriculture* 24: 201–214.

Sister, C., J. Wilson, and J. Wolch. 2007. The green visions plan for 21st century Southern California. 15. Park congestion and strategies to increase park equity. University of Southern California GIS Research Laboratory and Center for Sustainable Cities, Los Angeles.

Smith, K.E.C., and K.C. Jones. 2000. Particles and vegetation: Implications for the transfer of particle-bound organic contaminants to vegetation. *The Science of the Total Environment* 246: 207–236.

Smith, W.R., R.M. Farrar Jr., P.A. Murphy, J.L. Yeiser, R.S. Meldahl, and J.S. Kush. 1992. Crown and basal area relationships of open-grown southern pines for modeling competition and growth. *Canadian Journal of Forest Research* 22: 341–347.

Souch, C., and S. Grimmond. 2006. Applied climatology: Urban climate. *Progress in Physical Geography* 30: 270–279.

Spash, C.L., and A. Vatn. 2006. Transferring environmental value estimates: Issues and alternatives. *Ecological Economics* 60: 379–388.

Spiro, T.G., and W.M. Stigliani. 2003. *Chemistry of the Environment*, 2nd ed. Prentice Hall.

St. John, J.C., and W.L. Chameides. 1997. Climatology of ozone exceedences in the Atlanta metropolitan area: 1-hour vs 8-hour standard and the role of plume recirculation air pollution episodes. *Environmental Sciences & Technology* 31: 2797–2804.

Stallins, J.A. 2004. Characteristics of urban lightning hazards for Atlanta, Georgia. *Climatic Change* 66: 137–150.

Stallins, J.A., and M.L. Bentley. 2006. Urban lightning climatology and GIS: An analytical framework from the case study of Atlanta, Georgia. *Applied Geography* 26: 242–259.

Stallins, J.A., M.L. Bentley, and L.S. Rose. 2006. Cloud-to-ground flash patterns for Atlanta, Georgia (USA) from 1992 to 2003. *Climate Research* 30: 99–112.

Sullivan, W.C., F.E. Kuo, and S.F. Depooter. 2004. The fruit of urban nature: Vital neighborhood spaces. *Environment and Behavior* 36: 678–700.

Sun, H.L., M.C. Chou, and K.H. Lue. 2006. The relationship of air pollution to ED visits for asthma differ between children and adults. *American Journal of Emergency Medicine* 24: 709–713.

Sutherland, W.J., S. Armstrong-Brown, P.R. Armsworth, T. Brereton, J. Brickland, C.D. Campbell, D.E. Chamberlain, A.I. Cooke, N.K. Dulvy, N.R. Dusic, M. Fitton, R.P. Freckleton, H.C.J. Godfray, N. Grout, H.J. Harvey, C. Hedley, J.J. Hopkins, N.B. Kift, J. Kirby, W.E. Kunin, D.W. Macdonald, B. Marker, M. Naura, A.R. Neale, T. Oliver, D. Osborn, A.S. Pullin, M.E.A. Shardlow, D.A. Showler, P.L. Smith, R.J. Smithers, J.L. Solandt, J. Spencer, C.J. Spray, C.D. Thomas, J. Thompson, S.E. Webb, D.W. Yalden, and A.R. Watkinson. 2006. The identification of 100 ecological questions of high policy relevance in the UK. *Journal of Applied Ecology* 43: 617–627.

Sutton, P.C., and R. Costanza. 2002. Global estimates of market and non-market values derived from nighttime satellite imagery, land cover, and ecosystem service valuation. *Ecological Economics* 41: 509–527.

Svedsater, H. 2003. Economic valuation of the environment: How citizens make sense of contingent valuation questions. *Land Economics* 79: 122–135.

Swartz, J., P. Michaels, and R.E. Davis. 2005. Ozone: Unrealistic scenarios. *Environmental Health Perspectives* 113: A86–A87.

Taha, H., H. Akbari, and A. Rosenfeld. 1991. Heat island and oasis effects of vegetative canopies: Micro-meteorological field-measurements. *Theoretical and Applied Climatology* 44: 123–138.

Talbot, J.F., and R. Kaplan. 1984. Needs and fears: The response to trees and nature in the inner city. *Journal of Arboriculture* 10: 222–228.

Taylor, A.F., F.E. Kuo, and W.C. Sullivan. 2002. Views of nature and self-discipline: Evidence from inner city children. *Journal of Environmental Psychology* 22: 49–63.

Taylor, G.E., D.W. Johnson, and C.P. Andersen. 1994. Air pollution and forest ecosystems: A regional to global perspective. *Ecological Applications* 4: 662–689.

Thompson, C.W. 2002. Urban open space in the 21st century. *Landscape and Urban Planning* 60: 59–72.

Tolbert, P.E., J.A. Mulholland, D.L. Macintosh, F. Xu, D. Daniels, O.J. Devine, B.P. Carlin, M. Klein, J. Dorley, A.J. Butler, D.F. Nordenberg, H. Frumkin, P.B. Ryan, and M.C. White. 2000. Air quality and pediatric emergency room visits for asthma in Atlanta, Georgia. *American Journal of Epidemiology* 151: 798–810.

Troy, A.R., J.M. Grove, J.P.M. O'Neil-Dunne, S.T.A. Pickett, and M.L. Cadenasso. 2007. Predicting opportunities for greening and patterns of vegetation on private urban lands. *Environmental Management* 40: 394–412.

Vitousek, P.M., P.R. Ehrlich, A.H. Ehrlich, and P.A. Matson. 1986. Human appropriation of the products of photosynthesis. *BioScience* 36: 368–373.

Vogel, K.P., J.J. Brejda, D.T. Walters, and D.R. Buxton. 2002. Switchgrass biomass production in the Midwest USA: Harvest and nitrogen management. *Agronomy Journal* 94: 413–420.

Vose, J.M., G.J. Harvey, K.J. Elliott, and B.D. Clinton. 2003. Measuring and modeling tree stand level transpiration. In *Phytoremediation: Transformation and Control of Contaminants*. Eds. S.C., McCutcheon, and J.L. Schnoor, pp. 263–282. John Wiley and Sons.

Wackernagel, M., N.B. Schulz, D. Deumling, A.C. Linares, M. Jenkins, V. Kapos, C. Monfreda, J. Loh, N. Myers, R. Norgaard, and J. Randers. 2002. Tracking the ecological overshoot of the human economy. *Proceedings of the National Academy of Sciences*. 99: 9266–9271.

Walsh, C.J., T.D. Fletcher, and A.R. Ladson. 2005a. Stream restoration in urban catchments through redesigning stormwater systems: Looking to the catchment to save the stream. *Journal of the North American Benthological Society* 24: 690–705.

Walsh, C.J., A.H. Roy, J.W. Feminella, P.D. Cottingham, P.M. Groffman, and R.P. Morgan. 2005b. The urban stream syndrome: Current knowledge and the search for a cure. *Journal of the North American Benthological Society* 24: 706–723.

Warneke, C., S.A. McKeen, J.A. de Gouw, P.D. Goldan, W.C. Kuster, J.S. Holloway, E.J. Williams, B.M. Lerner, D.D. Parrish, M. Trainer, F.C. Fehsenfeld, S. Kato, E.L. Atlas, A. Baker, and D.R. Blake. 2007. Determination of urban volatile organic compound emission ratios and comparison with an emissions database. *Journal of Geophysical Research* 112: D10S47.

Weiss, P.T., J.S. Gulliver, and A.J. Erickson. 2007. Cost and pollutant removal of storm-water treatment practices. *Journal of Water Resources Planning and Management* 133: 218–229.

Weng, Q., D. Lu, and J. Schubring. 2004. Estimation of land surface temperature–vegetation abundance relationship for urban heat island studies. *Remote Sensing of Environment* 89: 467–483.

White, K.D. 1970. Fallowing, crop rotation, and crop yields in Roman times. *Agricultural History* 44: 281–290.

White, M.C., R.A. Etzel, W.D. Wilcox, and C. Lloyd. 1994. Exacerbations of childhood asthma and ozone pollution in Atlanta. *Environmental Research* 65: 56–68.

Wilson, J.S., M. Clay, E. Martin, D. Stuckey, and K. Vedder-Risch. 2003. Evaluating environmental influences of zoning in urban ecosystems with remote sensing. *Remote Sensing of Environment* 86: 303–321.

Wilson, W.G. 2000. *Simulating Ecological and Evolutionary Systems in C.* Cambridge University Press.

Wolch, J., J.P. Wilson, and J. Fehrenbach. 2005. Parks and park funding in Los Angeles: An equity-mapping analysis. *Urban Geography* 26: 4–35.

Wullschleger, S.D., F.C. Meinzer, and R.A. Vertessy. 1998. A review of whole-plant water use studies in trees. *Tree Physiology* 18: 499–512.

Zanobetti, A., and J. Schwartz. 2006. Air pollution and emergency admissions in Boston, MA. *Journal of Epidemiology and Community Health* 60: 890–895.

Ziska, L.H., D.E. Gebhard, D.A. Frenz, S. Faulkner, B.D. Singer, and J.G. Straka. 2003. Cities as harbingers of climate change: Common ragweed, urbanization, and public health. *Journal of Allergy and Clinical Immunology* 111: 290–295.

Index